Saufthorpe
Skendleby
Partney
Boothby
Ruins
Graby Hall
Welton
CANDLESHO
Raithby
Orby
Screnby
Candlesby
Enderby
Hundleby
Gunby
Ashby
SPILSBY
BOLINGBRKE
Halton
Holegate
Burgh
West Keal
East Keal
Bratoft
Great Steeping
H. Toynton All Saints
Irby
Little Steeping
NCBROKE
Firsby
L. Toynton
St. Peters
Bathing House and Tavern
Croft
Thorp Hall
Northolme
All Saints
WAINFLEET
St. Marys
Good Dike
White Cross
South Stream
Drain
Wrangle Bank
GIBRALTAR
Wainfleet Haven
Frisknev
Boston North Channel
Sibsey
Wrangle
Leak
Hall Ends
Leverton
Salt Marsh
Bay Hall
Bennington
IRBECK
Butterwick
Frieston
Bathing Houses
Frieston Shore
Fishtoft
BOSTON DEEPS
Roger
Boston South Channel
Hards House
WITHAM
Scale of Chains

To Mum

All my love

Mary

December 25/81

THE BOOK OF THE LINCOLNSHIRE SEASIDE

FRONT COVER: The sun rising over the flat sands and sometime tranquil sea.

Contemporary postcards of c1910.

THE BOOK OF THE LINCOLNSHIRE SEASIDE

The story of the coastline from the Humber to the Wash

BY

DAVID N. ROBINSON MSc

Published to mark the Centenary of the University of Nottingham

BARRACUDA BOOKS LIMITED
BUCKINGHAM, ENGLAND
MCMLXXXI

PUBLISHED BY BARRACUDA BOOKS LIMITED
BUCKINGHAM, ENGLAND
AND PRINTED BY
BURGESS & SON (ABINGDON) LIMITED
ABINGDON, ENGLAND

BOUND BY
KEMP HALL BINDERY LIMITED
OXFORD, ENGLAND

JACKET PRINTED BY
CHENEY & SONS LIMITED
BANBURY, OXON

LITHOGRAPHY BY
BICESTER PHOTO LITHO LIMITED
BICESTER, ENGLAND

DISPLAY SET IN BASKERVILLE
AND TEXT SET IN BASKERVILLE 11/12pt
BY SOUTH BUCKS PHOTOSETTERS LIMITED, ENGLAND

ISBN 0 86023 122 4

Contents

Introduction & Acknowledgements

It is particularly appropriate that this book should be written and published in the University of Nottingham's centenary year. When a student there just over 30 years ago, I was a member of one of the first research teams, led by Frank Barnes, to do survey work at Gibraltar Point. I was also fortunate, albeit briefly, to be influenced by Professor Swinnerton's unique style of encouraging lecture, then re-enforced by Dr (now Professor) Cuchlaine King's enthusiasm for beach studies. The storm-surge of 1953 stimulated more detailed investigation by the Geography Department of beach development and sand movement on the coast.

My initial researches attempted to show the relationship between the changing shape of the coastline and the physical forces acting upon it. More recently I have become fascinated with the story of 'man the beach-dweller' (homo sapiens littoralis) on the Lincolnshire seaside. In telling the many-sided story of that seaside, I have tried to weave together the geology and geomorphology with the economic and social history of a much under-rated stretch of Britain's coastline.

Walter White in 1865 may have been right in suggesting that 'romantic scenery must not be looked for on the Lincolnshire coast', although he did accept that 'the scene derives interest from its peculiarity'. The peculiar, that is special, features of dunes and marshes, sandy beaches and mud flats, with their complex of plants and animals, now draw many people to study courses mounted by my Department of the University (Adult Education) at the Gibraltar Point Field Station. These are most rewarding for tutors, if not also for students. A number on my own courses have produced short pieces of original work, and in particular I am grateful to Julia Rowley who catalogued my considerable collection of Lincolnshire beach pebbles.

In my attempts to convey an understanding of the methods of salt making I am especially indebted to the insight and inspiration of Mrs Ethel Rudkin, of coastal erosion and sea defence to Arthur Owen, and of the natural history of our seaside to the long-standing friendship of Ted Smith. Many individuals have assisted my accumulation of photographs over the years, but special mention should be made of Bill Hallgarth, the *Grimsby Evening Telegraph*, Winston Kime and Barrie Wilkinson. My ideas on salt making and smuggling have been superbly illustrated by Wally Day, and I am very grateful to Ken Redmore for redrawing my reconstructions of the Saxon and 13th century shorelines (the former owing much to the work of Brian Simmons) and the map of Wash reclamations.

Thanks are due to the Library Services in both Lincolnshire and Humberside, to the WEA Branches in the East Midlands and my own Department, and to the Lincolnshire & South Humberside Trust for Nature Conservation for the facilities which generated the subscribers who have made this book possible.

My wife sustained me when it seemed the writing would never be completed in time and straightened out the worst eccentricities of grammar, and Sylvia Turner willingly and cheerfully undertook the typing of the manuscript. For any errors that remain, or omissions that have been made, only I am to blame.

Dedication

For my parents
who first introduced me to the seaside

Foreword

By Professor Emeritus K. C. Edwards CBE, MA, PhD, FRGS

Following the success of *The Book of Louth*, David Robinson conceived the idea of writing the story of the Lincolnshire seaside to be published as part of the centenary celebrations of the University of Nottingham, his alma mater. Few authors know the coast as he does and its prehistory and history are unfolded with precision and zest.

I take pride in having him as one of my former students. He never wasted his time, either before or after graduation, and saw to it that his subjects—geography and geology—should be engaged in the understanding of man and his habitat. This he achieves through teaching, lecturing, fieldwork, editing and publishing and through secretaryship of his County Trust for Nature Conservation.

In this book he demonstrates the changes in human energies along the Lincolnshire coast against the inconstancy of the seaward margin. Conditions of place and time are compared and this is a true statement of environmentalism as it is now understood. May its popular expression by this well-acclaimed Lincolnshire scholar result in an increased understanding and enjoyment of the seaside.

K. C. Edwards.

Preface

By A. E. Smith, OBE, MA

Ours is not a scenically impressive coastline: there are no exciting cliffs, no rock pools full of life and colour, no picturesque fishing villages, no pine woods or heathlands in the immediate hinterland. Yet for those who come to know it the Lincolnshire seaside is a place of absorbing interest. Man and the sea between them have shaped this coastline: the story is one of erosion and floods, of sea defences and reclamation. By enterprise, ingenuity and tenacity man has won a living from these coastal lands—from salt-making in earlier times, from farming the rich marshlands through the centuries, to the more recent provision of entertainment for the holiday-maker. And whilst the seaside has provided clean air and enjoyment for vast numbers of Midlanders for more than a century, and whilst much of the sandy coast has been 'developed' for his purpose, wilder places have survived where the subtle beauty and abundant wildlife of sand-dunes, saltmarshes and seashores and something of the solitude that goes with them may still be enjoyed.

These are the main themes of David Robinson's fascinating and evocative book. He and I have shared many experiences in 30 years of friendship, not least a love of this coastline where we have both spent much of our lives. We have indeed been most closely associated in seeking to protect its natural beauty and interest and in helping more people to understand and appreciate it. It is a special pleasure for me, therefore, to commend this splendid book to the reader. You will find it a source of great interest and enjoyment.

Ted Smith

RIGHT: Postcard and Coat of Arms of Skegness,
and LEFT: Cleethorpes.

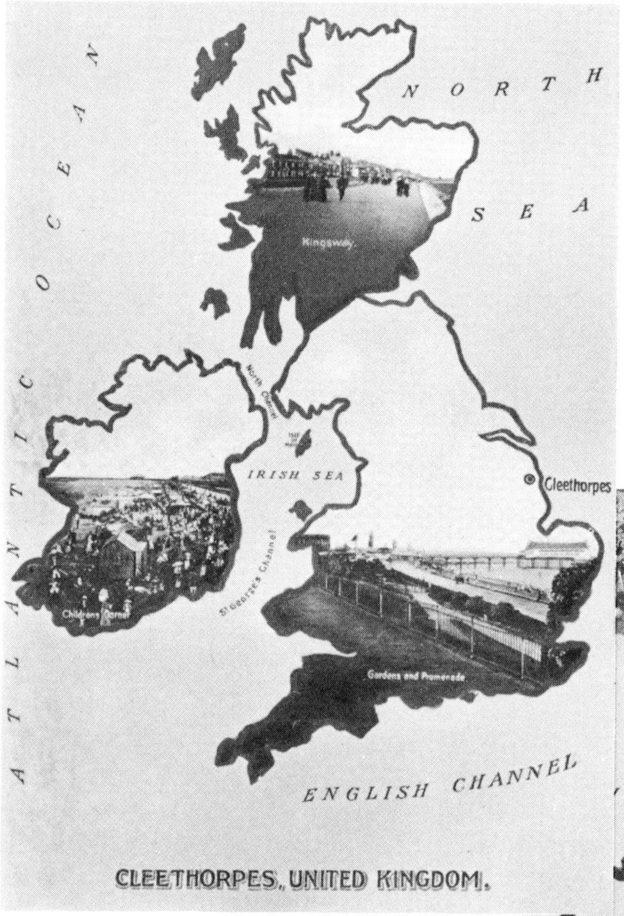

Pebbles on the Beach

The Lincolnshire seaside is sinking, sinking at the rate of about a foot a century, and has been doing for 1500 years. It has changed the map many times. Five mediaeval churches and two towns have been swallowed by the sea. In fact, were it not for today's solid sea defences, many square miles of the county would be flooded at the next high tide. Yet around 40,000 acres of new land have been won from the sea in the last 700 years. And a holiday industry has been built on the fact that the sea regularly washes the sandy skirt of Lincolnshire.

The Lincolnshire seaside of today is a by-product of the Ice Age. The pebbles on the beach are the clue. One of the first to recognise that the Lincolnshire seaside is one that has come in out of the cold was Joseph Correa de Serra.

The spring tide was at its lowest ebb as the boat put out from Sutton shore 'at half past twelve at noon' on 19 April 1796. The weatherbeaten fisherman had two unusual passengers—Sir Joseph Banks of Revesby, President of the Royal Society, and Dr Joseph Correa de Serra. 'It was a common report in Lincolnshire, that a large extent of moor, situated along its coast, and visible only in the lowest ebbs of the year, was chiefly composed of decaying trees.' Named 'clay huts' on Mitchell's coastal sailing chart of 1765, this was the 'submarine' or submerged forest they had come to see. The fisherman told them that similar 'moors' were to be found along the whole coast, from Grimsby to Skegness, particularly off Addlethorpe and Mablethorpe.

Setting foot on the 'islets' they found the remains of some of the trees still standing on their roots, with the trunks scattered on the ground, among which they distinguished birch, fir and oak. The trunks, branches and roots of the decayed trees were flattened. 'The soil to which the trees are attached,' they noted, 'is a soft greasy clay; but, for many inches above its surface the soil is entirely composed of rotten leaves.' What they were looking at, as also observed by A. J. Jukes-Brown for the Geological Survey in 1881 and H. H. Swinnerton of the (then) University College, Nottingham in the 1930s, is the post-glacial forest which grew between 9,000 and 6,000 years ago and the peat which strangled it. Many of the trees of that forest must have been of lordly proportions, for some of the stumps are as much as five feet in diameter, and fallen trunks 40 feet long and four feet wide have been seen lying in the peat. Between the 'islets' on which the forest occurred were channels four to twelve feet deep. De Serra concluded that 'little doubt can be entertained of the moory islets of Sutton being part of this extensive subterraneous stratum, which, by some inroad of the sea, has been stripped of its covering of soil.'

What apparently Banks and de Serra did not look at were the pebbles on the beach. It is these, together with the submerged forest, which are the clues to the early history of the Lincolnshire seaside. But the significance of the stones was not recognised until the 19th century, particularly when William Buckland championed the theory of glaciation in the 1830s, as immortalised in a poetic 'Dialogue between Dr Buckland and a Rocky Boulder':

'Has some strong torrent driven thee from afar,
Or has thou ridden on an icy car?'

By 1875, when the Religious Tract Society published *A Book for the Seaside*, the Victorian visitor to the Lincolnshire seaside was taking an interest in the 'countless number of pebbles, long since rounded by the action of the waves', but without appreciating just how they came to be on the beaches—nor yet the origin of beach sand itself.

An examination of these pebbles (particularly those of sufficient size to fit the palm of the hand) which are usually best seen at places like Huttoft Bank, Wolla Bank, Vickers Point and Gibraltar Point, reveals an enormous variety of rock types, quite apart from the fascinating shapes, colours and textures. Among those to be found are igneous rocks like pink granite from near Aberdeen, basalts, gabbros, syenites and volcanic ashes from the Central Lowlands of Scotland, dolerite 'bluestones' from the Cheviot Hills, and rhomb porphory from Norway (with its distinctive rhomboid crystals); metamorphic rocks such as garnet schists, Lewisian gneiss and quartzites from the Northern Highlands of Scotland; and sedimentary rocks like dark red sandstones from Central Scotland, millstone grit and grey limestone (on which ice scratches are best seen) from the northern Pennines, carbonaceous siltstones from the Durham coalfield and calcareous mudstones from Robin Hood's Bay, together with chalk and flints, the latter of almost every conceivable shape, including flakes which show clear signs of having been fashioned by man. There are fossils—sea urchins, crinoid (sea lily) stems, bivalve shells, bullet-shaped belemnites, sponges, corals, coiled ammonites, the 'devil's toe-nail' (gryphaea), the bark impressions of coal-age trees, and even the occasional dinosaur vertebra. And there is also the thrill of searching for the semi-precious stones—agate, carnelian, jasper and amber. How did it get there?

The story begins around 100,000 years ago when the coastline was the eastern edge of the Lincolnshire Wolds—an impressive line of chalk cliffs from Barton to Candlesby. Rising to nearly 300 feet and with a flinty beach at the base, they would have presented a picture rather like the white cliffs of Dover. The temperate climate and high sea level was one phase of the Ice Age—the Ipswichian Interglacial. As it became increasingly cold in the Last (Devensian) Glaciation, which began 75,000 years ago, early Stone Age Man, who had walked the cliff tops and pebbly foreshore, left for warmer climes. Sea level fell as water became locked up in snow and ice, packing down into glaciers forced outwards from the mountains of Scotland and the hills of northern England. Lumps of rock, plucked from the many different kinds over which the glaciers passed, were frozen into the ice as it coalesced into massive ice sheets, moving relentlessly south, and joined by ice from Scandinavia. The main advance of the ice over-rode the chalk cliffs and penetrated into the Fenland basin as far as Stickney and Sibsey. As the climate ameliorated, the ice wasted away, depositing an uneven mass of clay mixed with boulders, pebbles and gravels. But this was only a temporary phase, followed by re-advances of the ice-sheet, or periods of standstill or halt (that is when the rate of melting of the ice edge matched the push of the ice from the north). Over a period of perhaps 10,000 years (from 25,000 years ago) the result was to leave hummocky ridges of glacial deposits in the Middle Marsh from Aby to Hogsthorpe and from Killingholme to Cleethorpes (to form the cliffs there). Associated with them, but now off the present coastline, were extensive spreads of summer icemelt sands and gravels which are now the Docking Shoal, the Saltfleet, Theddlethorpe and Protector Overfalls (the latter a significant name), and the Haile Sand Flat.

In the immediate post-glacial period (called the Flandrian) east Lincolnshire was much higher than it is now, with the sea far away across the Northsealand and a tundra-type vegetation, including birch, growing on the extensive glacial debris. The (Boreal) period from 9000 to 5500 BC was drier and gradually it became warmer; first pine was dominant but was overtaken by oak to become the characteristic woodland, extending from the Wolds into the Northsealand. And Stone Age Man returned (Mesolithic and Neolithic) to hunt the deer and other animals in the woods. By the end of the period the sea had refilled the North Sea basin, and separated Britain from the continent, although in Lincolnshire the shoreline was still about 10 miles to the east.

Around 5500 BC sudden changes occurred. While the climate remained warm, it also became wetter—the Atlantic period. There was a rapid sinking of land relative to sea level, and consequent deterioration of land drainage. These two factors combined led to the growth of peat bogs, 2-3 feet thick, particularly in the hollows of the hummocky boulder clay. It was this peat which smothered the roots of the trees and entombed both the stumps and the trunks as they fell. This is the submerged forest that de Serra described. Examination of pollen grains in the peat shows that lime, alder, hazel, beech and elm also grew in the forests.

As Bronze Age immigrants began to arrive from the continent about 1800 BC, the mild, moist climate gave way to drier conditions, but as the land went on sinking the sea encroached more. It flooded through the offshore banks of sand and gravel between Holderness and North Norfolk and into the shallow Fenland basin, covering the peat with layers of clay, 8-10 feet thick, on which grew saltmarsh plants like those we know today. The effect of the spread of the sea was to leave the offshore banks as a series of low islands some 15 feet above sea level, with an extensive sheltered tidal lagoon behind. It is also likely that what are now the Dowsing Overfalls and Inner Dowsing began to form at this time, as a sand bar across the main entrance to the lagoon. However, land sinking not only ceased towards the end of the Bronze Age but there was a slight rise of about two feet, remaining more or less stable for about 1,000 years. A wide zone either side of the present coastline and the inner Fenland basin was no longer reached by high tides. The marsh became fresh; reeds and rushes replaced the saltmarsh. About 600-500 BC there was a sudden deterioration in climate to cool and wet, with freshwater floods and accumulation of a thin peat. By now Iron Age man had arrived and was making salt.

In Roman times the coast was a wide and gently shelving zone with freshwater marsh or peat (particularly in the interior Fens) grading into saltmarsh or wide sand and mud flats, intersected by long tidal creeks and river estuaries (some, like the Witham and Lud, having more than one outlet) flowing into the lagoon. The Wash was much broader and shallower than now, as the land stood at least 20 feet higher, but with extensive saltmarsh islets. Yet the present Outmarsh and Fenlands were lower, as they lacked the deposits formed later on. The so-called 'Roman' bank was not built then; it was a speculative invention of 17th and 18th century antiquaries Sir William Dugdale and Dr William Stukeley. Roman roads reached the shoreline at a number of places, notably near Burgh le Marsh, with a shore fort near Skegness and a signal station under the present Ingoldmells Point. If the Romans did have sea defence works on the central Lincolnshire coast, they have long since been lost to the sea. What they almost certainly had was a ferry across the Wash lagoon, from somewhere near Skegness to Thornham near Brancaster in Norfolk.

Towards the end of Roman occupation there was renewed land subsidence and inundation by the sea, starting slowly in the 2nd century and becoming more rapid in the 4th and 5th centuries. Within the Fenland lagoon silting up was rapid, with the formation of an irregular bar of coarse silt, to above the high water mark of ordinary tides, on the seaward edge of the marshy islets from Wainfleet to Boston and in a broad arc round to Spalding and Wisbech, covering many Romano-British settlement sites. Between the outfalls (outlets) of the Lymn and the Lud, the sea deposited some seven feet of sloppy clays, containing shells which indicate that the area was invaded by each tide. These clays lapped round the boulder clay hummocks, again covering Roman sites, and forming the flat Outmarsh. Between the Lud and the boulder clay promontories of Humberston and Cleethorpes a smaller silt bar developed.

Lincolnshire's Saxon shore at the 'dawn' of the Dark Ages may have had an air of desolation and decay, but its offshore sandbanks and inviting tidal estuaries had distinct similarities to those of north-west Europe, from which the land-hungry colonising Anglo-Saxons began to arrive from about 450. The sloppy clays and shallow creeks did not attract, but the Humber did, as evidenced by Humberston (Umber-stan or Humber Stone, which was one of the bluestone glacial erratics) and a line of personal name villages from Healing to Killingholme—all on the edge of the boulder clay, above possible flood level. East of Louth, Anglo-Saxon villages were also established well back from the coastline on the edge of the Middle Marsh from Covenham to Gayton; those settlers entered Lincolnshire along the tidal Lud. Immigrants navigated the main lagoon entrances by the Dowsing sand bars, and founded shoreline settlements around the 7th century on the silt bar in Holland ('the high land'). These were near deep tidal creeks and estuaries in the arc from Leverton, Butterwick and Freiston, Stenning, Quadring (Haefringas—dwelling in the mud) and Donington round Bicker Haven, Spalding (Spaldas—the dwellers by the gulf, the Welland), Whaplode (eel-pout stream), Holbeach (deep river), Fleet (tidal stream) and Gedney (Gydda's island) to Lutton (pool town) with its hamlet Sutton.

The late 8th and early 9th centuries not only ushered in a wetter period, but also saw fierce seasonal raids by the Danes, followed by an organised army. Having established the Danelaw, the 'retired' soldiers sent for their families and became another wave of colonists. The distribution of their villages (with the 'by' ending) and outlying farmsteads (the 'thorpe' ending) shows how they occupied the rich pastures of the Lincolnshire marshes not occupied by earlier settlers. Some were established on the hummocks of boulder clay which protruded through the layers of silt, their distinct island-like nature indicated by their names—Hannah, Helsey, Huttoft and Listoft. On the Wash coast the Danes filled in gaps between existing settlements as at Fishtoft and Wigtoft, Wrangle (crooked creek) and Skirbeck (bright brook), Algakirk and Gosberkirk (now Gosberton).

The coastline between Mablethorpe and Skegness (Skeggi's ness, or headland where the Roman shore fort had been) was up to a mile or so further east than it is now. The sinking of the land had slowed to something like its present rate (six inches to a foot a century). The scavenging of the glacial sands of the offshore banks and the sea floor by the rising sea level began to produce beaches and sandhills—the first natural sea defences in the coming conflict between Canute and Neptune.

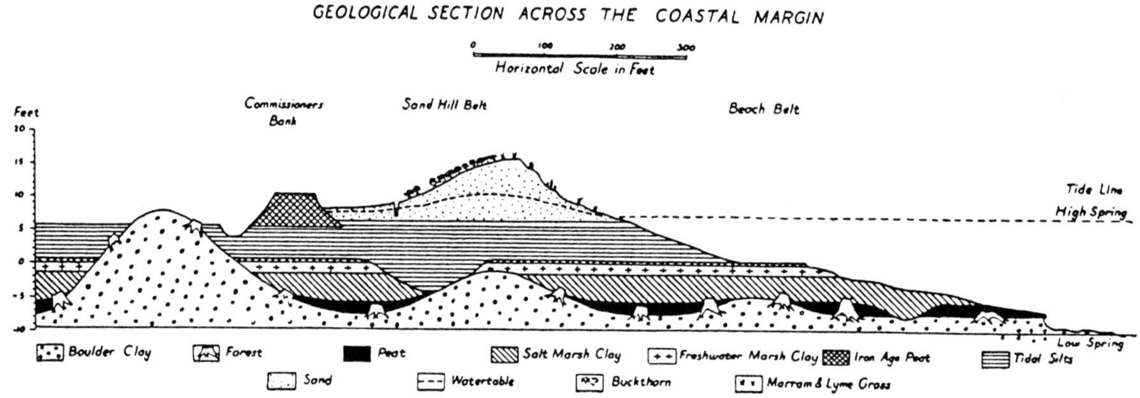

Geological cross-section of the coastline. (LNU)

ABOVE: Submerged forest at Trusthorpe seen at low water spring tide, September 1952. (GFM) BELOW: Tree bole and roots in peat and boulder clay on the shore at Huttoft.

Some Lincolnshire beach pebbles: ABOVE: Ammonite (left), gryphaea (centre), peat stained flint tool only 1½″ long (right); CENTRE: 'Bubbly' basalt from Scotland (left), rhomb porphory lava from Norway (right), a garnet schist from Scotland (below); BELOW: an agate (left), dinosaur vertebra (right).

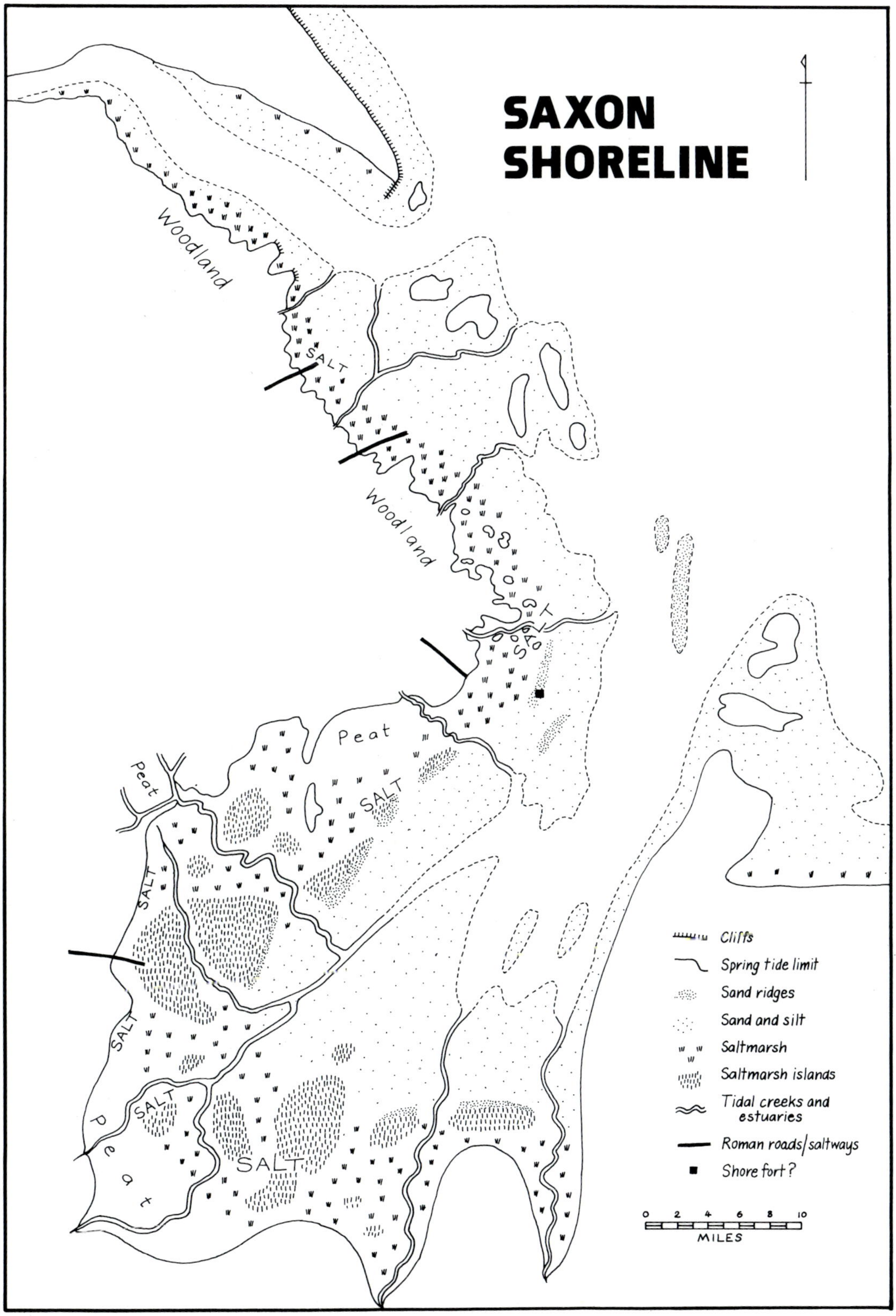
SAXON
SHORELINE
Woodland
SALT
Woodland
SALT
Peat
Peat
SALT
SALT
SALT
SALT
SALT
Peat
Cliffs
Spring tide limit
Sand ridges
Sand and silt
Saltmarsh
Saltmarsh islands
Tidal creeks and estuaries
Roman roads/saltways
Shore fort?
0 2 4 6 8 10
MILES

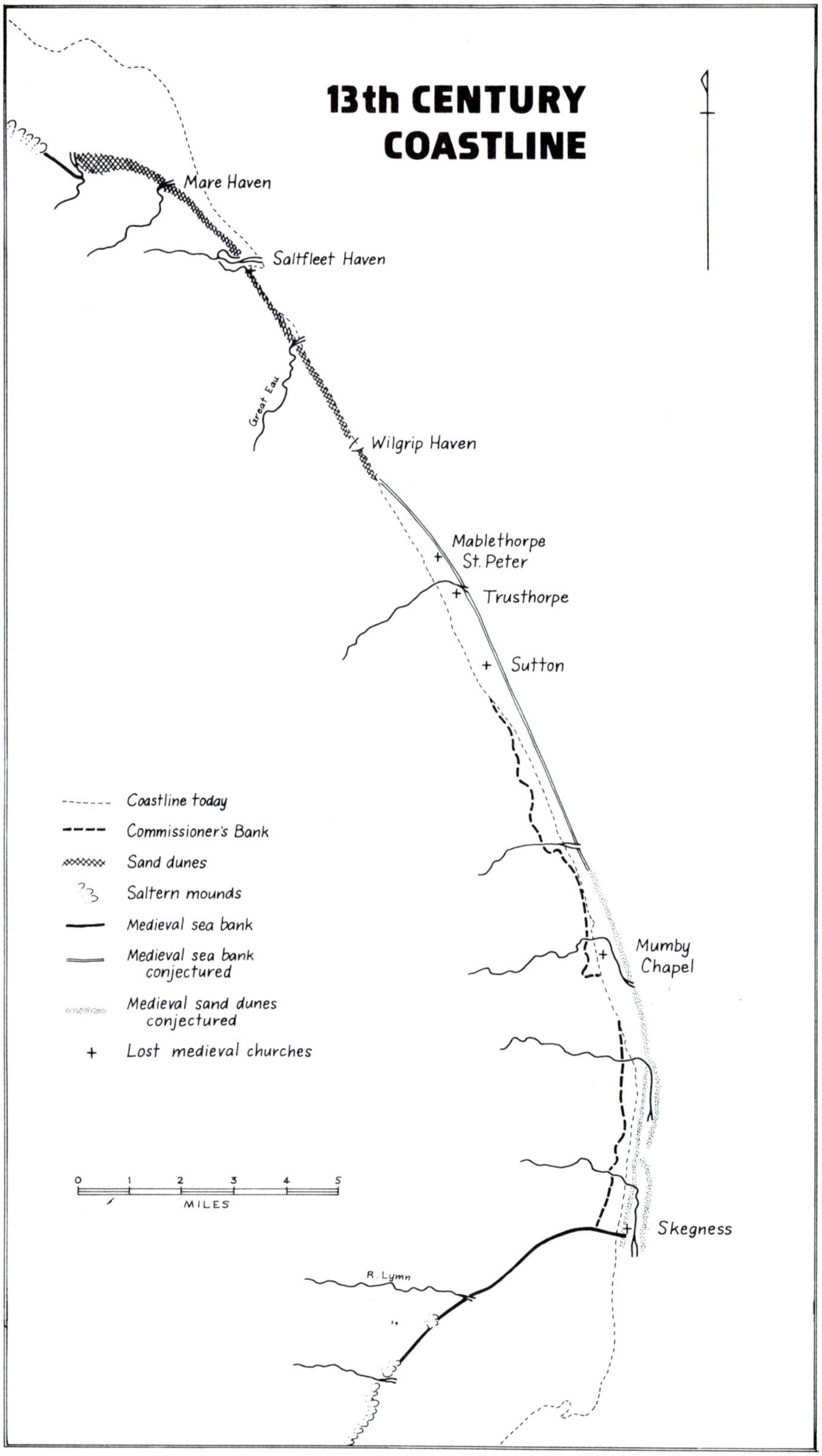

13th CENTURY COASTLINE
Mare Haven
Saltfleet Haven
Great Eau
Wilgrip Haven
Mablethorpe St. Peter
Trusthorpe
Sutton
Mumby Chapel
Skegness
R. Lymn
Coastline today
Commissioner's Bank
Sand dunes
Saltern mounds
Medieval sea bank
Medieval sea bank conjectured
Medieval sand dunes conjectured
Lost medieval churches
0
1
2
3
4
5
MILES

The Lost Towns

The sea looseth and gaineth considerable in this county . . .

Christopher Merret (1696)

The four to five centuries of post-Roman subsidence had been a period of adjustment to changes on the shoreline. As well as sand from the barrier shoals and the seafloor, fine silt from the erosion of Holderness in Yorkshire was washed in to balance the effects of subsidence. Where there was plenty of new sand on the still gently shelving foreshore, as at Somercotes and Skegness, it was spread by longshore drift (wave action carrying and depositing material along the shore), in ridges, with the first 'ness' at Skegness beginning to form. Onshore winds began to form low dunes, and there were also strips of saltmarsh. As the coast was possibly ten feet higher than now, there could be grazing on the Outmarsh and fringing saltmarshes, first in summer (la Sumerette in Fulstow, and Somercotes) and then more permanently. Where there was saltmaking in the north-east the mounds could be settled and joined with short sections of bank. There the coastal villages developed as daughter settlements: North Coates from North Thoresby, Foulestowemersch (Marsh Chapel) from Fulstow, Grainthorpe from Covenham/ Yarburgh, Conisholme from Alvingham, and Somercotes/Skidbrooke from Cockerington. Further south, permanent communities in the shelter of dunes or near stream banks developed similarly: the Saltfleetbys from Grimoldby/Manby, the Theddlethorpes from Gayton, the Mablethorpes from Strubby, Sutton in the Marsh from Hagnaby/Markby, and Mumby Chapel (Chapel St Leonards) from Mumby.

As the Outmarsh settlements became permanent, artificial banks (the sea dyke or hafdic) were built in gaps in the dune system, which sheltered settlements like Ingoldmells, (Ingiald's meols or sandhills) and across the marshes between the Skegness promontory and the Wainfleet saltworks (later called the 'Roman' bank which now carries the A52 road, and also Skegness High Street). The upkeep of the sea dykes was the responsibility of the dikereeves; each village annually elected two. The dunes and banks were crossed by drainage outfalls; the Lud had three—at Suine (in Grainthorpe), Mare (on the North/South Somercotes boundary) and Salfluet (Saltfleet Haven); there was the Great or Withern Eau and Wilgrip Haven (at Crook Bank) in Theddlethorpe, and other drain outfalls further south.

In the more sheltered Wash, where deposits more than balanced subsidence, there were both marshy and sandy shorelines, used for grazing (with banks for permanency) and saltmaking respectively. The combined effect was to reshape the Saxon shoreline. Funnel or trumpet-shaped havens and estuaries became constricted by saltern mounds and small 'inliks' or intakes of saltmarsh; Leake, Frampton and Kirton Havens totally filled up. The Nene estuary was particularly large as it also carried one distributary of the Great Ouse; its diversion to the King's Lynn outfall before 1292 caused changes in the pattern of the saltmarsh build-up. Between Pinchbeck and Lutton reclamations of large areas of 'neuland' was on a communal basis, with the Common Sea Dyke built over a mile out in the 13th century. They included one in 1286 by Thomas de Holbeach, who made an agreement with the villages of Holbeach and Whaplode for a common sea-dyke, together with arrangements for drainage gutters and sluice. This was in front of Holbeach Bank between Saracen's Head and Coney Garth (the abbot of Crowland's rabbit warren, which implies some build up of sand on the bank). It was the last reclamation of any consequence for three and a half centuries.

Man was already on the losing side. There had been the abnormally high tide of 1253 when the sea flowed up the River Lud as far as Alvingham, and floods in Holland. There had also been floods in Holland the previous century, requiring the repair of sea banks. In 1176 'the sea burst forth beyond its bounds . . . engulfed many men and herds', and two years later 'the submersion of Holland, by an inundation of the sea', but these did not deter them from further reclamations. And there were floods in 1248, 1250 and 1251. Clearly it was a stormy period with a number of tidal surges. Not only did these put at risk the land behind the piecemeal sea banks, low sandhills and saltern maures or tofts (hillocks), but the accompanying wave action severely eroded the protecting offshore shoals (already 6-10 feet lower, due to rising sea level).

In the second series of major floods, the worst was on the night of New Year's Day 1287 when the church of St Peter, Mablethorpe was 'rent asunder by the waves of the sea', and in Holland 'the whole country for the most part turned into a standling poole, so that an intolerable multitude of men, women and children were overwhelmed with the water, especially in the towne of Bostone'. The bank built the year before by the men of Holbeach and Whaplode was breached, and repairs were not completed until about a century later.

The surges of 1287 and 1288 finally overwhelmed the offshore barrier shoals. The debris of shingle and sand from the Overfalls was driven on to the adjoining shoreline, forming a broad storm beach, on which wind-blown sand formed dunes up to 50 feet high, from Somercotes narrowing towards Mablethorpe, and a smaller one from Cleethorpes to Humberston Wadhouse. Finer silt was swept into the Wash, to be kept in circulation like a huge sedimentation tank. The Lincolnshire seaside no longer looked out on to a lagoon, but to the open sea. The new shape of the coast was not adjusted to the marine forces of tide, wave and wind acting upon it directly. The process of adjustment—either resisted or aided by man—is still proceeding. The convex curve of the coast between Saltfleet and Skegness, with its mixture of dunes and sea banks, laid it open to erosion. Here Neptune always seemed to be in the ascendant.

This continued into the 14th century. Sea banks at Pinchbeck were broken by 'tempestuous waves' in 1323, and the sea broke through banks at Mablethorpe in August 1335, drowning sheep and crops. The increased amount of sand and silt on the coastline adversely affected havens and outfalls. Mare Haven had fallen into disuse by the end of the 12th century, the 'port' of North Coates had silted up by 1274, and by 1347 the Great Eau had been diverted into Saltfleet Haven to assist in keeping it clear for trade in corn, beans and fish (and in 1359 the haven provided two ships and 49 men for Edward III's invasion of Brittany).

At Ingoldmells, seabanks were in disrepair in 1365; by 1404 Cattysacre was devastated by the sea, and the parish suffered flooding in 1421 and 1422. Eight years later there was breaching and flooding along much of the coast from Mablethorpe to Skegness (where 'land was submerged by flow and reflow of the sea') and the Commissioners of Sewers for the 'province of Lindsey' were directed to take steps to safeguard the sea coasts and marshes and to impress as many workmen as they deemed necessary 'considering the great and urgent necessity to hasten the same'. In 1443 Thomas Fitzwilliam (lord of Mablethorpe manor) was exempted from offices and services 'in consideration of his loss of land in Mablerthorp through the irruption of the sea and of his costs in repairing the coasts', and in 1485 rents to the lords of Ingoldmells were reduced because part of the land chargeable was 'swallowed up by the sea'. A survey by the Commissioners in 1503 showed Mablethorpe and Skegness to be 'in very great danger of the sea', and 14 years later, despite cleaning dikes and repairing banks in Skegness, the sea 'rushed at last over the barriers which had been raised on this level shore, and recovered his ancient possession'. But much worse was to come.

In 1526 the 'church and a great part of the parish (of Skegness) was submerged'. By the time Leland arrived (about 1540) at the 'sumtyme great haven toune' which had traded in Baltic timber, the destruction was complete. Mr Paynelle told him 'he could prove that there was ons an haven and a toune waullid having also a castelle' (the remains of the Roman shore fort?) 'The old toune is clene consumid and eten up with the se. At low water appere yet manifest tokens of old buildings'—about three quarters of a mile out to sea. There had been the church where John Tuttoft's father had been to school and from which Christopher Woolthorpe's father had helped to carry stones to build a new church (St Clement's) well inland. What appears to have happened is that the neck of the ness of dunes which formed a promontory SSE from near Ingoldmells (the 'South Meles' or 'long rigg' which sheltered the haven and saltmarshes) had given way. Around the same time the hamlets of West Meales and the East Meales disappeared, together with a cony warren on the sea bank and Warren House, where the warrener had entertained gentlemen resorting to the bowling green on the marsh. The ness broke up, began to reform and extend south again, and in about two centuries had formed the first Gibraltar Point. Meanwhile the Commissioners followed their policy of building a new bank further back—the 'Skegness and Winthorpe country bank' (the 'Roman' bank north from Lumley Square in Skegness)—begun in 1568 but still not complete in 1574. Even so, 140 acres of marsh were 'lost unto the sea irrecoverably' in 1616 (or so they thought). However, about 1670 Lord Castleton enclosed outer marshes in front of the country bank with the 'Green Bank' (between High Street and North Shore Road), outside which were the 'Outslaps' or muddy foreshore (where later were the dunes of 'The Jungle').

A fate similar to that of Skegness overtook the church of St Peter, 'Mawplethrop' and the greater part of its parish in the early 1540s when they were 'overflown with water in the sea' although part of the church ruins could be seen NE of the present pullover as late as the 1870s. Apparently no serious attempt was made to create an adequate sea defence, and over a century later it was 'only Banks or Hills of a small sand called Meals'. A little further to the north at Saltfleet the 'chapel of the cross of the sands' dedicated to St Katherine, near the then mouth of the haven, was also lost.

The greatest disaster of the 16th century was the high tide, accompanied by tempest of wind and rain, on 5 October 1571. (The event was immortalised in the title of Boston poetess Jean Ingelow's poem 'The High Tide on the Coast of Lincolnshire, 1571'—'The olde sea wall (he cried) is downe, The rising tide comes on apace'—but the details are those of the 1810 flood.) Two churches and part of the villages were lost—at Sutton and 'Mumby Chappell' (Chapel St Leonards) where a new sea bank was under construction but incomplete; consequently 'the church was wholly overthrown except the steeple' and 'the whole town was lost, except three houses'. Ships were driven onshore all along the coast; at Mumby Chapel 'a shippe was driven upon an house' and the sailors only saved themselves by climbing on top of it. (The chapel was partly rebuilt and finally disappeared when the present church was built, again well inland, in 1794.) From the Humber to the Wash there were massive losses of stock—20,000 cattle and 4,000 sheep, including 1,100 belonging to Mr Spicer, between Humberston and Grimsby, which drowned with their shepherd as 'the water breake fiercely' on the saltmarshes. Within a century or so of this disaster there was little to be seen of Sutton church, even at low water. In 1637 the inhabitants there had petitioned the Privy Council, complaining that the Commissioners had failed to provide a new bank, while the town 'hath been divers years and yet is in great danger to be destroyed by the rage and violence of the sea'. The mediaeval church at Trusthorpe was probably destroyed in the 1571 flood (the first building on the present site dates from 1606).

The method of sea bank repair, or the construction of a new bank (usually further inland, as the Commissioners recognised the inevitability of further erosion) was to dig clay from in front, or behind, the bank line and move it by horse and sled to where the 'bankmen' or 'spademen' were working. The main season for this work was May and early June (between seed-time and haymaking) as the demand for labour was considerable. A new bank might require a work force of over 300. There was also carpentry work in building sluice-gates (gouts or gotes), piles and 'jetty work' (groynes), timber being brought in by sea (for instance from Tattershall via the River Witham and Boston). Brushwood 'hedges' or kidding were set in the seaward side of banks to trap blown sand.

The costs of the work in the Lindsey coastal parishes were met by assessing 'acresilver', a rate calculated on the number of acres a man owned. When work on a drainage outfall was required, or there was a serious emergency due to damage to sea banks, inland parishes also had to pay a levy. The instructions of the Commissioners of Sewers to the dikereeves for the repair of two-thirds of a mile of sea bank at Theddlethorpe in 1571 were that it was to be 13 or 14 feet high when wet 'so that when it is thoroughly settled it remain 12 foot plum high, and to batter to the seaward 30 foot from the top to the skirt without any hollowness; and to be 6 foot broad in the top and to batter to the land 10 foot . . . And to be made only with clay, and all sand to be clean rid at the foundation, that clay may be laid to clay'. This work was financed by a rate of 12d on every acre in Theddlethorpe and 5d on every acre in all the towns within its levy. The impoverishment of coastal parishes by land losses and the expenses of bank repair is reflected in the mean and jumbled churches at Mablethorpe, and at Sutton (described in 1789 as 'a most wretched church of stud and clay with a wooden ruinous steeple'), and the lack of chancels at Ingoldmells and Addlethorpe where parishioners petitioned in 1665 for their removal on account of 'very much good ground . . . lately lost by the great irruptions of the sea'.

In Holland the cost of bank repair was met by agistment or joycement (rates of payment on pasture or arable holdings—a form of tithe). The length of bank would be measured and staked out to mark the lengths for which individuals were responsible. When danger threatened, the whole community was there, 'when the north-easter is blowing above, the spring tide roaring outside, the brimming tide-way lapping up to the dyke top, or flying over in sheets of spray; when round the one fatal thread which is trickling over the dyke, or worse, through some forgotten rat's hole in the side, hundreds of men are clustered . . . The sheep have been driven off the land below; the cattle stand ranged shivering on the higher dykes inland: they will be moved by punts if the worst befall, but a hundred spades, wielded by practised hands, cannot stop that tiny rat hole. The trickle becomes a rush, the rush a roaring waterfall. The dyke top trembles—gives. The men make efforts, desperate, dangerous . . . with faggots, hurdles, sedge turf; but the bank will break . . . through the breach pours a roaring salt cataract'. To protect sea banks the pasturing of stock on them was prohibited; at Skegness in 1574 this extended to 'all marsh grounds, waste grounds and pasture' between the old and new banks. Rabbits were also a problem, and in 1765 an Act was passed 'for preventing the Damage done to Sea Banks, within the County of Lincoln, by the breeding Conies therein'. At Ingoldmells there were constant fines for cutting 'sines' (marram grass) which bound together the sand of the protecting dunes.

OPPOSITE ABOVE: The poor condition of Sutton church early this century reflects the impoverishment of the parish by having to maintain sea defences. CENTRE: The chancel is still missing from Ingoldmells church. BELOW: Rectangular clay pits dug to repair mediaeval sea banks exposed on the shore at Ingoldmells.

Saxton's map of the Lincolnshire seaside 1607.

Robert Morden's map of the Lincolnshire seaside 1695.

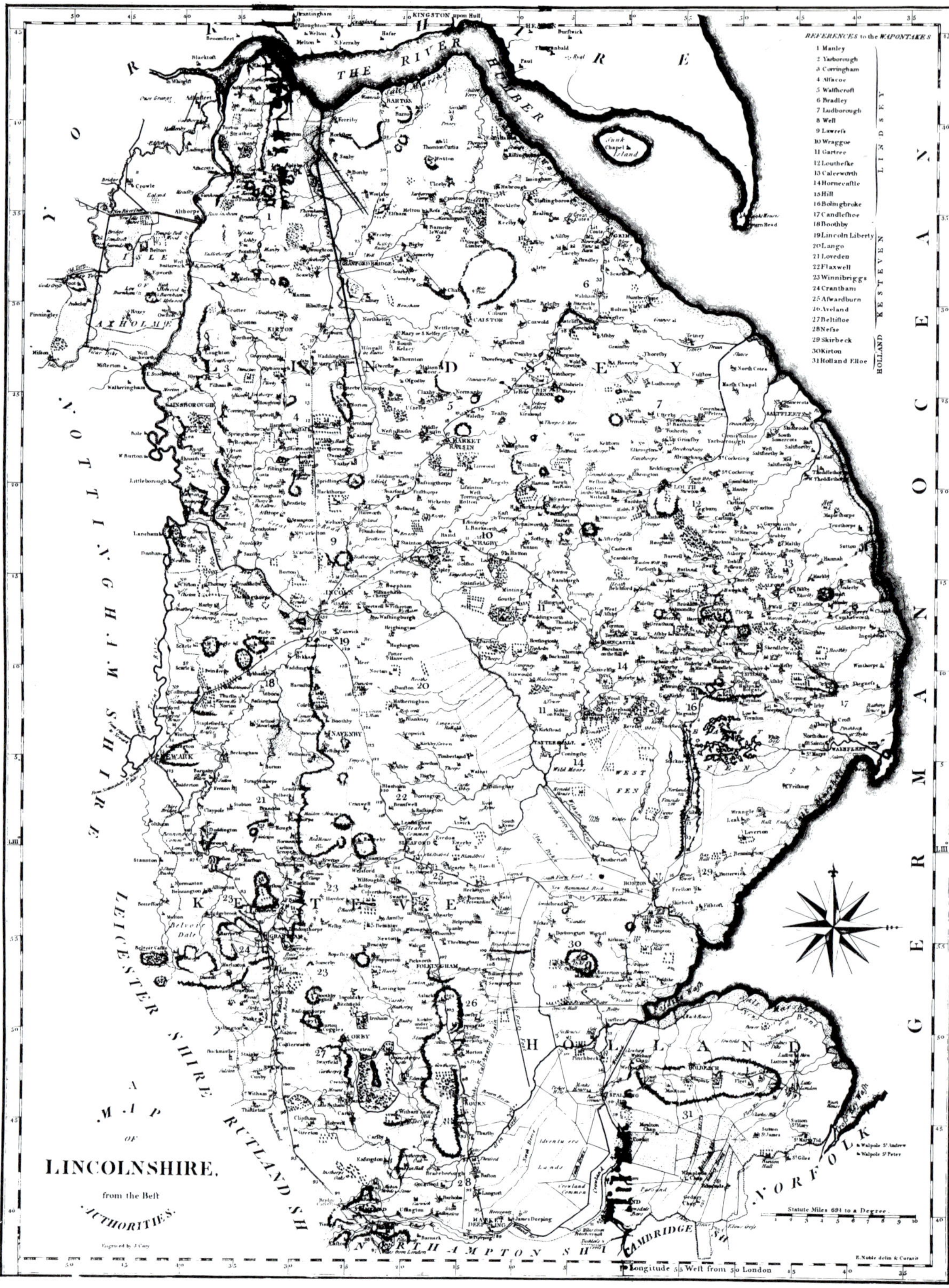

John Cary's map of the Lincolnshire seaside 1789.

The New Lands

Although there were intermittent inundations in the 17th century, as at Hogsthorpe Jackfish Bank between Chapel and Ingoldmells, when the 'Out End Dyke grounds' were flooded in 1637; at Skegness in 1639 (dunes broken) and at Mablethorpe in 1645; Wilgrip Haven silted up, and there was accelerated erosion of the Cleethorpes cliff (with erratic blocks as large as churches falling out, it is said) and consequent silting of Grimsby Haven, the century also saw changes of a different kind. Man had not been defeated. Where salt had been made for seven centuries the mounds of debris had grown outwards, but more significantly there had been considerable growth of saltmarsh in three sheltered areas—on the north-east shoreline within the Humber, in front of Croft and Wainfleet sheltered by the extending Gibraltar Point, and within the inner Wash between the Welland and Nene estuaries (where the effect of tidal flow and ebb always drives silt ashore).

On the north-east coast there was a wide zone of deposition, as far as the present low water mark before the end of the 16th century, with 40-50 feet thickness of sand and silt over the post-glacial clays. The saltmarshes here are called 'fitties', an old Norse word for salt pastures of the foreshore ('alway is drouned at Spring Tydes'). In Tetney parish there had been a Westseadyke or Inseabank, with a 'Shipdock', but an Eastseabank 'no bigger than a plough furrow' was hardly adequate to withstand the 1571 flood. By 1576 it was a more substantial feature linking the saltcoteholmes to enable Newton, Lescastles and South Marshes to be recovered from the sea—amounting to 1,000 acres, and valued at 5s an acre. In front of the new banks, Newton Fittyes (420 acres) was reserved for the use of Vincent Sheffield, Edward Lacon and Thomas Skegness; Castle, Fleet and South Marsh Fittyes, totalling 200 acres and regularly overflowed by the tide (worth only 2d an acre) were common to all tenants. In the adjoining parish of North Coates, 205 acres of old saltcotes had been embanked and turned into arable land in 1569.

In front of the Croft 'Country Bank' the first reclamations for over 300 years of what had become an extensive saltmarsh (over 1,500 acres) were made—around 1555 at the Skegness end and in 1589 at the Wainfleet end (between the Old Steeping River and the Toynton Beck which drained the East Fen through Wainfleet and formed its haven). That haven was decaying in 1543 due to silting, and a new sluice was erected further down the haven by 1608. In Wainfleet parish at least 140 acres had been enclosed out of 1,100 acres of saltmarsh; 300 acres were reclaimed in 1620 and by 1638 a further 560 acres were ripe for reclamation. Reclamations were continuing apace south of Skegness—400 acres in 1618 and a further 400 acres by 1627. In 1641, between 200 and 300 acres were enclosed in Friskney by building a new bank a hundred yards out from the edge of the Tofts. Here there were many small owners because of the strip division of the foreshore for saltmaking. Even with the rapid growth of saltmarsh, new banks could be in danger from high tides, and the new Friskney bank was breached in two places in 1686. Conditions seemed to have stabilised somewhat on the central Lincolnshire coast, as reclamations (or perhaps more accurately recovery of land) had been achieved by 1613 at

Huttoft (524 acres), Anderby (477 acres) and even Mumby Chappell (196 acres).

As a result of law suits in the 1560s about reclaimed marsh, it was established that 'if . . . the sea abandon the . . . salt shore, whereby it become of value, then the King's right prevails'. To assert this right, James I offered certain villages in Holland, where saltmarsh in the Wash had grown to its greatest extent (up to 3 miles wide), the chance to buy their common marsh. In 1613 only the men of Moulton and Long Sutton took up the offer. Moulton bought 724 acres for £105 2s 4d and Long Sutton 700 acres for £758 6s 8d. This gave them the right to any further increments of marsh. Grants of other saltmarshes, at an annual rent, were offered to outsiders on condition that they dike and embank at their own expense, including the Crown's share (one fifth and usually the best part) and that for the commoners. Thus in 1615 marshes in Wigtoft, Moulton, Whaplode, Holbeach and Tydd St Mary were granted by James I to Charles Glenmand and John Walcot of London as nominees of the Earl of Argyll, and marshes in Gedney and Sutton were sold to Sir Henry Wotton and Sir Edward Dymoke. The commoners challenged these grants as it reduced their pasturage (in Wigtoft from 412 to 120 acres).

The expanses of common marsh included an estimated 2,500 acres in Moulton and 2,000 acres in Gedney. However 'the numberinge of the acres is a verie uncertaine thinge . . . for there wilbe some tymes a hundrethe acres of marshe ground: and within three howers space the best of it wilbe overflowed with the sea above six foote deepe'. They were used for pasturing sheep and horses, with shepherds' huts at the upper end of the marsh. After spring tides it was the practice to wait for several showers to wash away sand and some of the salt. The low marsh was worth only 1d or ½d an acre, but the high marsh varied from 2d to 6d and 1s (in Kirton and Frampton), with the very best grazing in Long Sutton at 2s an acre. The financial incentives to enclosures were too great to be put off by threats of dispossession by the King. Just over 1,000 acres were enclosed in Holbeach between 1580 and 1620. Sir William Welby sold 580 acres of reclaimed marsh there for over £5 an acre.

The really big reclamations got under way in 1632 when the Dutch engineer Sir Cornelius Vermuyden embanked 1,120 acres in Tydd St Mary in the Nene estuary. In 1640 Charles I granted 6,670 acres in Sutton and Lutton to the Duke at Lennox at £300 a year, with power to enclose. By 1660 a massive 17,374 acres in Gedney, Whaplode, Holbeach and Moulton had been reclaimed by Adventurers, and by the same date the whole of Bicker Haven had been cut off from the sea. In fact a traveller crossing the Welland washes from Lynn to Spalding in 1634 'found not so much water (in the river) as would drown a mouse'. It was a totally new coastline, which more than compensated for three centuries of losses on the central Lincolnshire coast. Moreover, because of the rising sea level over that period, the new land surface in these reclamations was three feet higher than that inside the Common Sea Dyke.

Meanwhile on the north-east coast London courtier Endymion Porter (Gentleman to the Royal Bedchamber) was granted about 2,000 acres of commoners' marsh in North and South Somercotes in 1632. The local inhabitants refused to supply lodging and food for the imported labour force, but embanking and draining were completed by 1636. This ambitious project entailed a sea bank some five miles long, part of which forms the present headland from Donna Nook (supposed to take its name from the first body washed ashore there after a storm) to Stonebridge. Although a first crop of oats and barley was harvested in 1638, the commoners took advantage of the Civil War to throw down the banks and retake possession. By 1638, a bank had been raised extending to the present headland of Horseshoe Point, enclosing 600 acres in the parishes of North Coates and Marsh Chapel. A new sea bank was built round Porter's Marsh in 1702, but it was not long before John Porter was in trouble with the Commissioners of Sewers for stocking conies in the banks. The land in this enclosure had been divided between Porter, Edward Tottie (of London), Sir Gervase Scrope (lord of the manor) and the commoners (who received a quarter—still known as Fivehundred Acres).

The story of the 18th century is one of continuing conflict with the sea but man was winning. In 1720 William Newland (an appropriate name) reclaimed 1,332 acres on the west side of the Nene estuary but went bankrupt and the land was bought by Guy's Hospital. On 16 February 1735 a boisterous north-west wind and spring tide overflowed a mile of Ingoldmells Bank, and most of the land south of Dydick Bank in Ingoldmells and Addlethorpe parishes was flooded for three weeks. The following year there was 'a great and sudden inundation of the sea' at Anderby, and there were 'prodigious strong tides' in February 1750 and December 1757. A great tide in August 1744 broke the bank round Porter's Marsh in Somercotes in a number of places.

In 1747 Guy's Hospital reclaimed a further 762 acres in the Nene estuary, but the Moulton bank was breached in 1765 drowning 2,000 sheep, 13 horses and seven beast. Also in 1765 work had started on the Louth Navigation Canal, the first section from Tetney Lock to Firebeacon opening two years later, and that to Louth in 1770, the same year that 400 acres of saltmarsh in Grainthorpe parish were added to Porter's Marsh. Under the Tetney Enclosure Act of 1774, about 690 acres of the North and South Fittys were reclaimed, on both sides of the new canal. Parts of the sea bank (40 ft wide at the base, 6 ft at the top and 7½ ft high) were constructed of chalk, brought by barge from Hessle on the north side of the Humber estuary. Engineers employed by the Tetney Drainage and Enclosure Commissioners during construction in 1778-79 were John Hudson of Louth and Thomas Hogard of Spalding. While this work was in progress the sea made a 'great breach of the banks' in 1777 at Sutton, where eight acres had been lost by 1789. But also in 1789 there was more reclamation in Croft and Wainfleet marshes, and in 1795 Anthony's Bank was completed round the Wadhouse Enclosure in Humberston.

On the Holland coast, the area of most rapid saltmarsh growth was still between the Welland and Nene estuaries. An Act of 1793 formed a Trust of ten private landowners (the largest were Lord Boston, the Earl of Buckingham, N. Garland, W. Drake, and M. Dayrell) to build the South Holland Embankment, or Sixteen Mile Bank, from Wrag Marsh Farm above Fosdyke Washway to Boat Mere Creek in Gedney. The bank was 63½ ft wide at the base, 11 ft high with an outer slope of 1 in 4 flagged with 2½ inch sods and sown with rye-grass, and cross banks were built in case of breaches. There were major sluices for the Holbeach, Dawsmere and Boatmere Creeks and for the Old Fleet Haven. The area reclaimed was 4,595 acres, at a cost of about £10 an acre. The work was completed in 1811, despite damage the previous year.

That was caused by the terrible floods of 10 November 1810. Although no land was lost permanently to the sea, it was a disaster of considerable proportions. The cause was an easterly gale throughout the whole day increasing to hurricane force as the evening high tide was due. From the Humber to Boston Deeps some 40 vessels were wrecked, including ships from Shields, Blythe, Sunderland, Newcastle, Whitby and Stettin, *La Minerva* (a fire ship lately captured from the Danes), nine Boston fishing smacks, and a barge which drifted over the sea bank near the Scalp into the pastures. With every succeeding tide, mangled corpses were washed up. Every sea bank on the shores of the Wash suffered, not least the new banks made the previous year to reclaim a narrow strip of marsh from Wainfleet to Friskney and Wrangle; at Freiston scarcely a vestige was left, while the old 13th century bank behind was also broken in many places. The old sea banks here and at Skirbeck, Wyberton, Frampton and Fosdyke were of insufficient height to withstand this surge, which passed over them and scoured breaches from the land side. At Plummer's Hotel, just behind the old seabank at Freiston Shore, the great bow window of the dining room on the first floor was forced from the building and carried several fields away. Among flotsam thrown up by the tide were two casks of spirits at Gedney; these were found by two labourers who helped themselves so plentifully that one died in consequence.

The effect of the easterlygale, piling up water in the South-east corner of the Wash, was to create a tidal bore, or eagre, in the winding outfall of the Witham. This is what Jean Ingelow described in the poem based on this event: 'A mighty eygre reared his crest, And uppe the Lindis raging sped'. By 7 o'clock the tide had risen four inches higher than that of 1807, the water was 2ft 8½ ins deep in Boston Stump (St Botolph's church) 'And boats, adrift if yonder towne Go sailing uppe the market place'. In fact the tide was 4ft 7½ins above an ordinary spring tide, that is from 6 to 10 feet above the surface of the land. At Fosdyke the tide overflowed the bank so suddenly that a servant maid of Mr J. Birkett was surrounded by the sea while milking the cows in a pasture, and drowned within sight of the farmhouse. She was 'my sonne's wife, Elizabeth' of the poem.

The damage to agriculture was massive: farm buildings and cottages destroyed, an estimated 15,000 sheep drowned as well as cattle and horses, implements washed away, corn stacks wrecked (although a hay stack of Mr Day at Freiston was moved bodily 400 yards without damage), apart from the problem of getting rid of salt water and repairing banks. Total damage was over £28,000. Engineer John Rennie was called in to examine the banks, and estimated that repairs costing £33,000 were required to give a bank top two feet above the 1810 tide level.

During the 19th century the final, familiar shape of the Lincolnshire coastline as a blunted peninsula took shape, and the outfalls of the Wash rivers—Witham, Welland and Nene—were controlled. There were also floods. These had become inevitable on the central part of the coast if the banks gave way, because most of the Outmarsh is three feet below ordinary high water mark. The mark of the 1837 flood on Winthorpe church is 8ft 9ins above ground level. Bank repair became increasingly expensive: after the extraordinary tide of 1883, sea defence works costing £20,000 were required at Sutton and Trusthorpe.

On the north-east coast reclamations became possible east of Porter's Marsh and the Somercotes Warren: 300 acres of Donnanook Newmarsh about 1840, 250 acres of the Saltfleet New Enclosure in 1854 (effected by 'fixing a line of fascines to arrest the sand which blows off the shore') when Saltfleet Haven was straightened with training walls, and the 125-acre strip of Thimbleby's Enclosure about 1863. On the other side of Porter's Marsh, 320 acres of saltmarsh were bought in 1841 from the Earl of Yarborough by Henry Pye of Louth, solicitor and Commissioner of Sewers, and his partner Thomas Waite (who later withdrew). The 1¾ mile bank and sluice were constructed in four months in 1843 at a cost of £3,000. In 1855 Pye built himself a marine residence just behind the new sea bank; a removeable brick in the bathroom wall gave him a view of the Donna Nook Lifeboat Station. However, in 1868 he was in severe financial difficulties and fled the country—rowing a boat (despite an artificial leg) down Somercotes Haven and out into the Humber, where he boarded a passing ship. Other reclamations were 450 acres of North Coates Fitties in 1856 to create a new headland at North Coates Point, and the big fitty embayment of 446 acres at the head of which was Grainthorpe Haven (where coal had been imported, hence Coal Shore Lane there) in 1858. Neptune's assault on the boulder clay cliffs at Cleethorpes was halted in 1883-85 and 1903-06 by the construction of promenades.

The last part of Croft Marsh behind Gibraltar Point was embanked about 1812 (and Wainfleet Haven improved under the Steeping River Act of 1885), but there were no further enclosures along the coast towards the Witham. Minor improvements had been made to the Witham outfall in 1829 and 1865 (with some small reclamations) but the New Cut to Clay Hole (under the 1880 Act) was not completed until 1885 (at a cost of £107,668 for 2½ miles of works and the excavation of 2m cubic yards of materials). This not only gave deeper water and saved 1½ miles on the upstream passage to Boston Dock (opened 15 December 1884), but it had the effect of stimulating new saltmarsh growth along the shores to the north.

Attempts to deal with the defective state of the Welland outfall commenced with the 1794 Act, but improvements only extended as far as the Fosdyke crossing, where at low tide in 1835 there were only a few inches of water. Under the Fosdyke Bridge and Embankment Act of 1811 'for making a public way over the bare sands' to the appropriately named Hope & Anchor Inn on Moulton Marsh bank, work began the following year and was complete by 1815. A bridge of oak, designed by John Rennie, had two opening sections for the passage of vessels, and the half-mile bank carrying a 40-ft road, was 12 feet high. Below the bridge 2¼ miles of training walls, of thorn faggots 6 ft by 3 ft weighted with clay or sods, were constructed to half-tide level in 1837-38; the effect of resultant scouring was to lower the river bed by 7 feet. When works were begun, the foreshore in Frampton and Moulton was all sand, but it had become grassed over by 1851. Reclamations either side of the outfall were made in 1813 and 1864 in Fosdyke, 1870 in Kirton, and 1875 in Moulton (this 400 acres undertaken by Lord Boston, Richard Jackson and Edgar Garland at a cost of £13 10s an acre).

After the 313-acre Shearcroft's Inclosure at the outfall of Lutton Leam in 1805-06, other reclamations on the South Holland coast were patchy. The 1833 bank protecting 900 acres reclaimed in Holbeach and Gedney was damaged by the gale of February 1835 and the contractor was ruined. The work was done again, by Johnson and Sturt in 1838 (called Bull Dog Bank, because when a bailiff brought a bulldog to assist in arresting one of the navvies, they killed the dog and buried it in the bank) and by Miss T. Steer in 1840-50 (when the area totalled 1,130 acres), and there were two reclamations in Gedney of 400 acres in 1865 and 360 acres in 1876.

The main change in the coastline was in Cross Keys Wash and the outfall of the Nene. Kinderley's Cut of 1773 had straightened the Nene down to Gunthorpe Sluice on the county boundary, but beyond there the outfall was constantly shifting, with quicksands. The 1½-mile crossing from Wash House (now the Bridge Hotel in Long Sutton) to Cross Keys House on the Norfolk bank was only fordable at low water. Drovers and cattle, horses and carriages had to be piloted by a guide (Charles Wigglesworth of Long Sutton, who died in 1840 at the age of 85, had been guide for 52 years). When John Rennie came to make plans for a new cut for the river, he was nearly drowned when his carriage was overtaken by the tide. Under the 1826 Cross Keys Bridge and Embankment Act and the 1827 Nene Outfall Act the works were begun simultaneously. A new 5 mile channel was cut (designed by Thomas Telford), with two lighthouses at Crab's Hole, and 1¾ miles of training walls out into the sands of the Wash. It took 1,200-1,500 men employed by contractors Jolliffe and Banks three years and cost £200,716. The final act in June 1830 was to have 300 men with 100 carts working six days and nights to seal off the old river course (sinking four or five barges into the gap as well). The resulting scour in the new Tycho Wing's Channel (named after the Duke of Bedford's agent) lowered the bed ten feet. The first ship, the 170-ton brig *Friends,* entered the channel on 19 June and passed the moveable cast iron centre span of the bridge (designed by Sir John Rennie) which had been completed two months earlier. The massive 1½-mile Cross Keys embankment (which also effected the reclamation of 1,500 acres of marsh) took 900 men and 260 horses only 26 weeks to build. The first vehicle to cross it was the Union coach from Norwich to Newark on 4 July 1831. The remaining areas of marsh in the Lincolnshire portion of the estuary were reclaimed at irregular intervals from 1846.

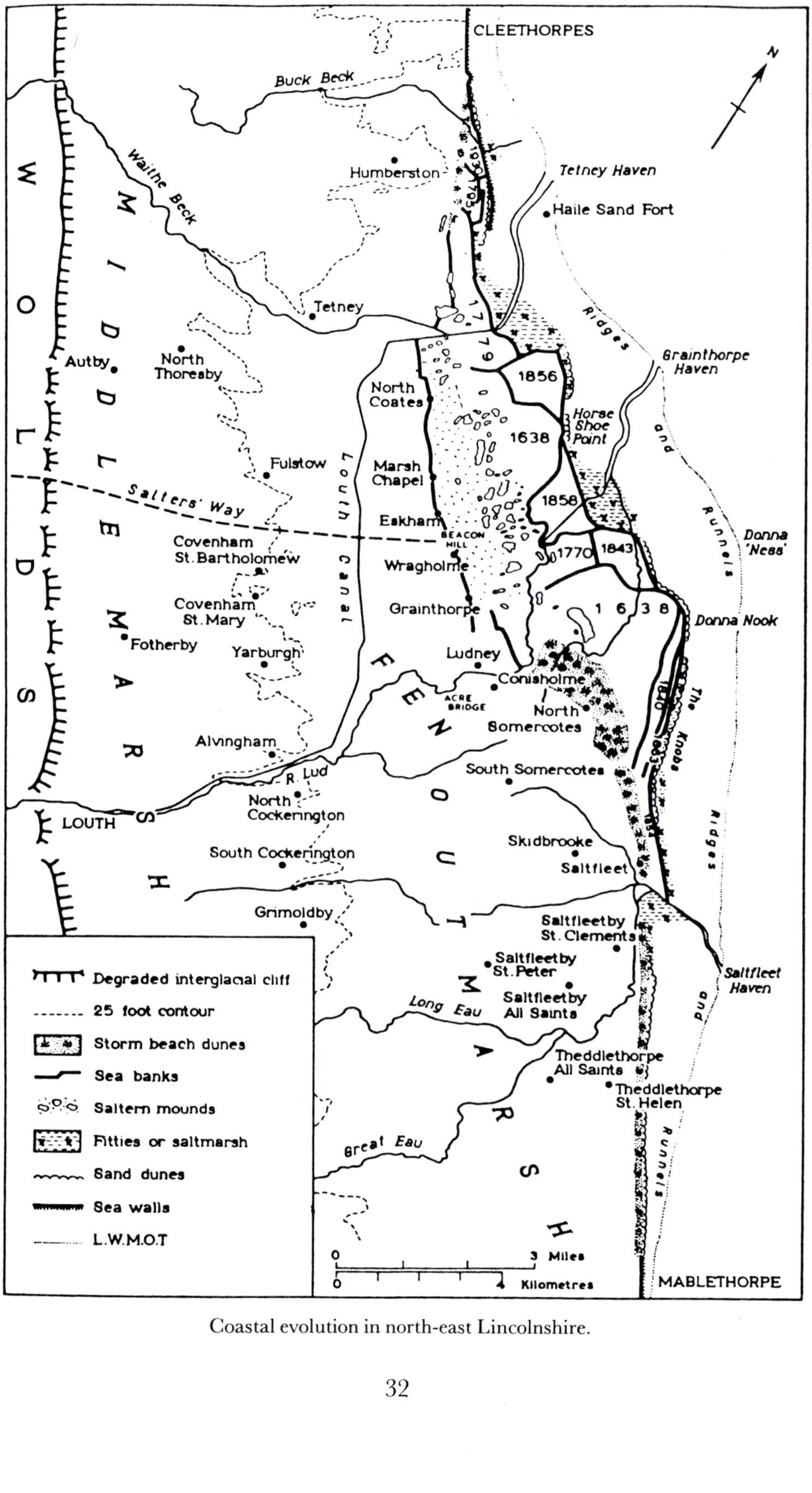

Coastal evolution in north-east Lincolnshire.

LEFT: Henry Pye. RIGHT: A gowt—pointed sea doors on a drain—at Saltfleet. CENTRE: Pye's Hall, North Somercotes. (BK) BELOW: Somercotes Haven.

ABOVE: The coastline at Saltfleet in 1860. BELOW: W. B. Thomas painting of the Witham outfall (New Cut completed 1885).

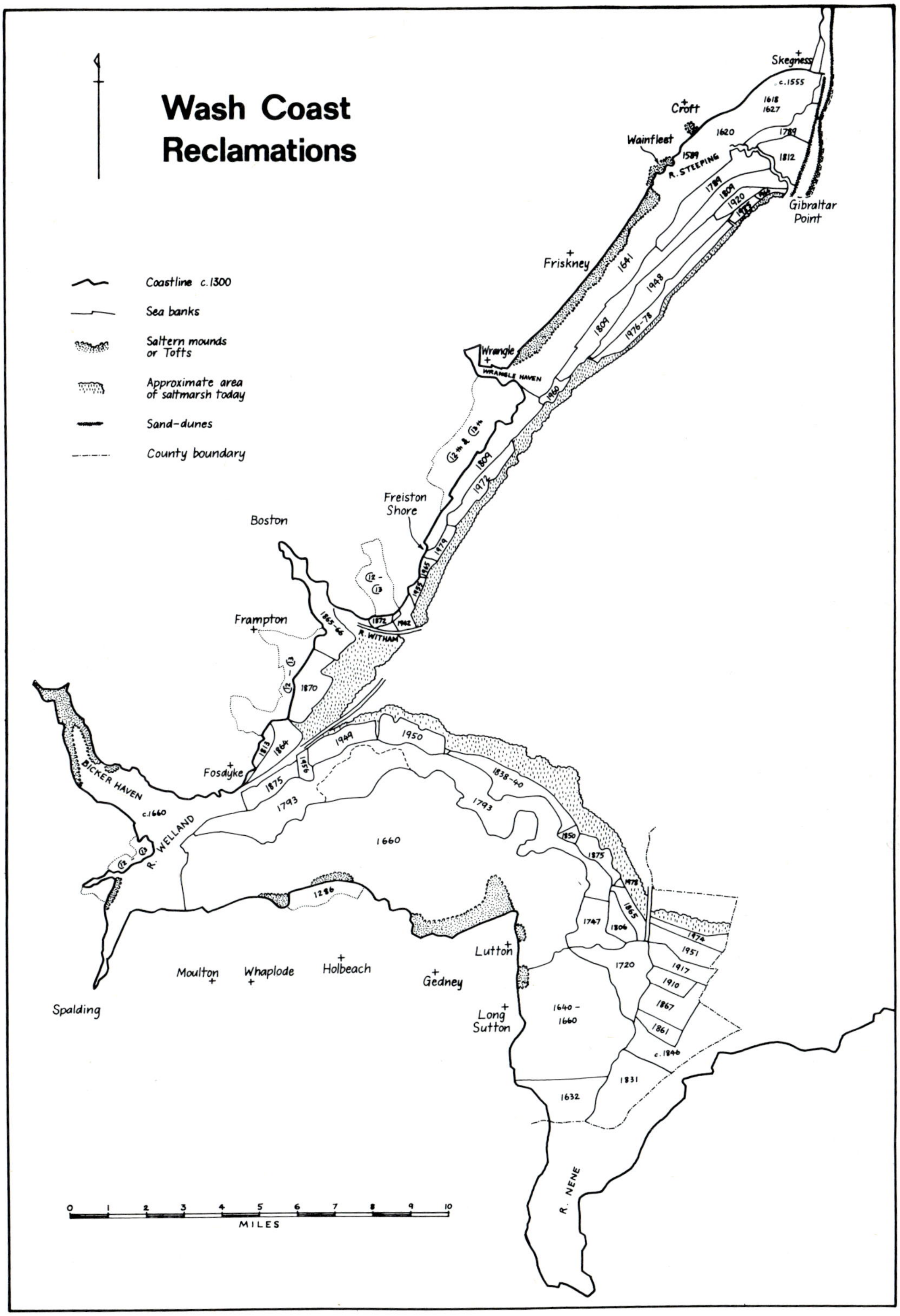
Wash Coast
Reclamations
Coastline c.1300
Sea banks
Saltern mounds or Tofts
Approximate area of saltmarsh today
Sand-dunes
County boundary
Skegness
Croft
Wainfleet
R. STEEPING
Gibraltar Point
Friskney
Wrangle
WRANGLE HAVEN
Freiston Shore
Boston
Frampton
R. WITHAM
Fosdyke
BICKER HAVEN
R. WELLAND
Lutton
Moulton
Whaplode
Holbeach
Gedney
Long Sutton
Spalding
R. NENE
MILES
0 1 2 3 4 5 6 7 8 9 10

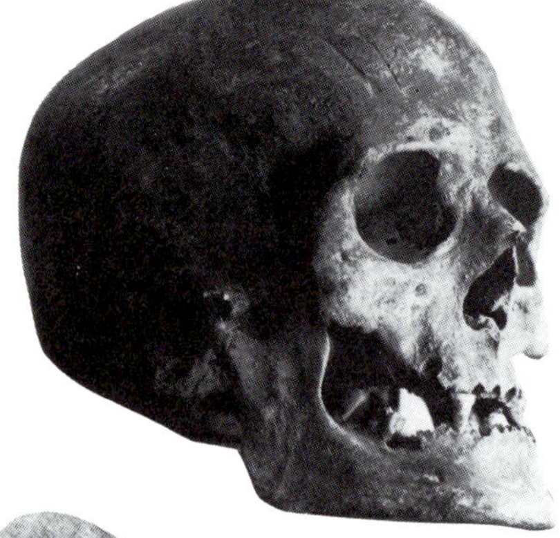

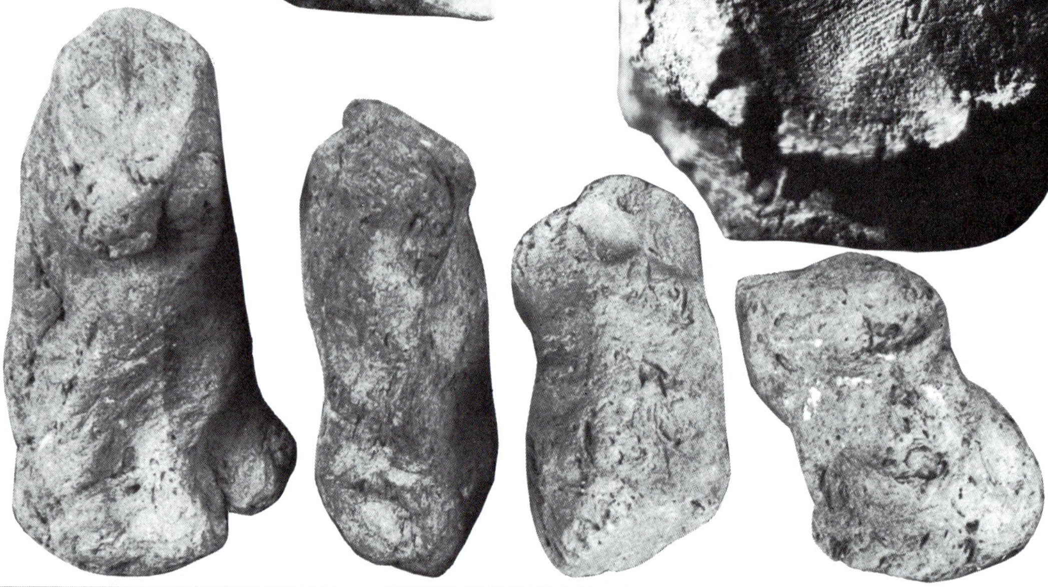

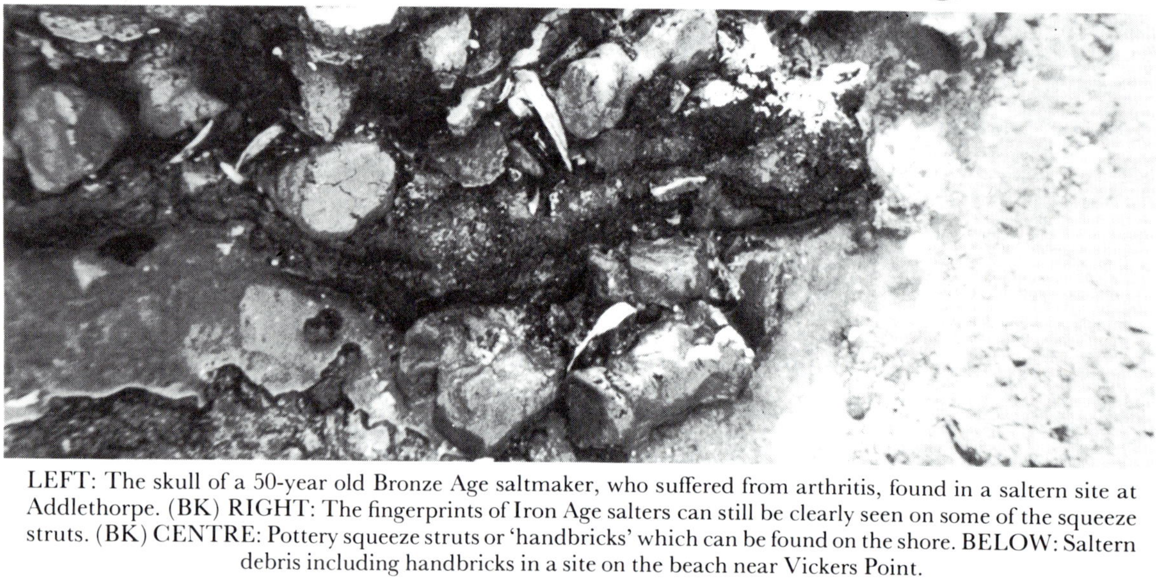

LEFT: The skull of a 50-year old Bronze Age saltmaker, who suffered from arthritis, found in a saltern site at Addlethorpe. (BK) RIGHT: The fingerprints of Iron Age salters can still be clearly seen on some of the squeeze struts. (BK) CENTRE: Pottery squeeze struts or 'handbricks' which can be found on the shore. BELOW: Saltern debris including handbricks in a site on the beach near Vickers Point.

Salt in Season

Teacher:	Ho, Salter! What doth thy craft profit us?
Salter:	My craft profiteth you all greatly; none of you enjoyeth his bait or dinner unless my craft befriend him.

(Aelfric's Colloquy)

When Edmund Cowper, salter of Marshchapel, died in 1559 the story of over 2,000 years of the making of salt on the Lincolnshire seashore was already drawing to a close. During that time it developed into a major industry, making Lincolnshire one of the country's leading producers. It started towards the end of the Bronze Age when the land began to rise slightly relative to sea level, generating freshwater marsh intersected by tidal creeks. The biggest spread of the industry was during the succeeding Iron Age. The workers in the prehistoric saltmaking industry occupied sites above high spring tide level in scrub country alongside streams flowing off the hills into a series of long tidal creeks through the saltmarshes, mud flats and sandbanks which extended into the sheltered tidal lagoon.

Sea water was scooped from the creeks at high tide, or from side channels which had been blocked off to allow the silt to settle and some natural evaporation to take place. It was then poured into shallow U-shaped pottery troughs about two feet long and nine inches across, quite light in weight and porous (the pottery containing chopped grass and chaff with grains of oats and ryegrass). Also scooped from the mud on the banks of the creek, or from the supply of sandy clay with vegetation debris from which the troughs and pots were made, were fistfuls of clay which were squeezed into struts, or handbricks, to support the trough over the fire in the hearth where the brine was to be evaporated. Sometimes small wedges of clay, squeezed between finger and thumb, were required to correctly balance the trough, or a tier of two or three troughs. The fire, having been lit with brushwood, was stoked with peat where it was available locally, or with turves. As salt crystals began to appear, more brine was poured in until there was sufficient salt to be chopped out when still damp. This was placed in other pottery vessels for the final drying process into convenient 'denominational' salt cakes for trading. The very nature of the processes involved meant that breakages were common, or even necessary to remove the salt (particularly where there had been a quick boiling to produce white, fine grain salt), and the site soon became littered with pottery fragments and unwanted baked strut supports, and these would help to build up the hearth for further firings. In time this created a low mound which, if there were a number of hearths together, could be 100 feet or more across. When sifting through this debris, it is interesting to grip the strut and try to determine whether it was a right-hander or a left-hander. An occasional fragment even preserves the actual finger or thumb print of its Iron Age maker.

The industry was widespread in both time and area, with very little relationship to the present coastline. The Wash then was a huge shallow bay of mudflats and marshy islands; the shoreline saltworks were in what is now Helpringham, Rippingale and Dowsby Fens, 15 miles from the present coast. One of the largest mounds, at Rippingale, grew to five feet in height and about 200 feet across, a sign of a considerable industry. On the basis of finds so far made (at least 100 saltern sites are known from this period) the greatest concentration of activity was in the present parishes of Hogsthorpe, Ingoldmells, Addlethorpe, Orby and Winthorpe. Here streams like the Burlands and Welton Becks meandered between low hummocks of boulder clay, which protruded through the marsh clays into tidal creeks, giving an ideal situation for salt-making. Most sites only come to light from beneath the post-Roman marsh clays when modern drainage or roadworks are undertaken, or by careful field walking. However, on the beach the shifting sands do sometimes reveal the burnt red clay of evaporating hearths, pottery debris and handbricks, usually around present mid-tide level.

Exactly how long the first main stage of saltmaking lasted is still unsure, but the sudden deterioration of climate, around 600-500 BC, to cool and wet with consequent freshwater floods undoubtedly ruined some early sites in the Hogsthorpe-Ingoldmells area. The saltmakers shifted their operations to a new location. With a further slight rise of land level towards the end of the Iron Age, there was an increase in saltmaking, and a spread eastwards with the newly developed shoreline. In this area of the lagoon shore the industry was taken over and became more organised and extensive during the Roman occupation; this was the area labelled Salinae (saltworks) by Ptolemy.

Within Metaris AEstuarium (the Roman Wash) the gradual accretion and growth of saltmarsh began to fill the inner part of the bay with low silty islands. It was on those marshy islands and at the tidal limits of creeks, particularly in the Whaplode-Holbeach-Fleet area, just west of Spalding and at Wrangle, that the new phase of saltmaking developed. On the western side of the bay the shoreline had moved a couple of miles east and new saltworks were established near Great Hale and Hacconby. The methods changed little from Iron Age times to the end of saltmaking at the close of the Roman occupation. All the sites were close to a fuel supply—the freshwater peat which had accumulated near the inner shoreline of the Metaris AEstuarium. Salt was produced for preservation of hides, mutton, fowl, fish and oysters, and was exported inland well beyond Lincolnshire.

Whether the decline of the Romano-British salt industry was gradual or sudden is uncertain, but undoubtedly its demise was hastened by a dramatic change in coastal conditions from the third century AD. With the subsidence of the shoreline, coarse beach deposits obliterated the salt workings, succeeded by up to seven feet of sloppy marsh clays. The old salt industry was dead. During the Dark Ages the new settlers from north-west Europe developed a new method of winning salt, but it is not until the Domesday record of 1086 that we can be sure of its location. However, what the survey did show was that the new industry was already well established and widespread, with some 260 saltpans in existence.

The Wash had decreased in area as the edges silted up permanently, leaving funnel-shaped estuaries and small havens. It was in the most sheltered of these inlets, where both sand and fine salt was deposited, that saltmaking was re-established. The most important in the 11th century were Bicker Haven with 67 saltpans (Donington, Bicker, and Stenning and Drayton in Swineshead), Leake Haven with 41, and there were 15 at Wainfleet. Saltmaking did not restart in the Ingoldmells-Chapel area as it became more exposed, with little or no silting, but a completely new area was developed on the north-east Lincolnshire shoreline, (although it is possible that there had been some Romano-British saltmaking just west of Conisholme and Marsh Chapel). Between Tetney and Saltfleet the rapid accretion of silt and sand eroded from Holderness and the still protecting offshore islands provided the right conditions. By 1086 there were 82 salterns, some belonging to villages on the Middle Marsh three or four miles away (North Thoresby, Fulstow, Covenham) or even up on the Wolds (Thorganby). Perhaps the most extreme example was Maidenwell, 350ft up on top of the Wolds; its solitary saltpan holding was probably between Saltfleet and Mablethorpe. Within the Humber estuary there was relatively short-lived saltmaking on the shoreline between East Halton and Stallingborough.

In the mediaeval industry, the basic requirement was an area of sandy silt on a gently shelving shoreline, the upper part of which was only covered by high spring tides. The north-east Lincolnshire coast was ideal, generating the greatest volume of industrial waste. The foreshore was divided into a series of carefully measured long narrow strips extending from the Hafdic (sea bank) down to below the limit of high spring tides. These fortnightly spring tides, occurring roughly two to two-and-a-half days after new and full moon, would be closely watched in the 1530s by Marsh Chapel salter Edmund Cowper, his father, also Edmund, and other family salters Codbert and Walter. They would be hoping for calm, warm and dry weather during the summer—the main season for saltmaking (the salters worked from 1 May to 10 November). If the weather remained dry, the upper foreshore could be left for a few days after the tides had fallen; then the sandy silt would have a high salt content. This natural drying-out process would be assisted by the fact that the shore had been harrowed before the high tides. The top inch or two, called moulde, or muldefang, was scraped up with a scoop-shaped device called a mould buyle (or bewell) drawn by oxen or horses (the Cowpers had more than 10 oxen and 30 horses between them) and dragged across the intervening saltmarsh to the maure or round ground where it was put in heaps and protected from the rain.

To separate the salt from the moulde, either a clay trench was dug or a wooden hoppe used with a filter bed of straw, and into this the moulde was shovelled. Water which had been kept from the high tides in a sump or pond was poured over the moulde, the salt from which was dissolved and the extra strong brine trickled into the bryne fat (brine vat), with a hen's egg floating in it to test the density. When the egg began to sink, the salt in the moulde was exhausted and it was thrown out. The accumulation of this industrial waste increased the height (up to 15 feet) and area of the maures. The super-saturated brine was evaporated slowly in round clay pans 15 feet across and 2½ feet deep at the centre, placed in a depression of similar size packed with peat, and later in leaden saltpans over a peat or turf fire in the saltcote or boiling house. Towards the end of the industry outcrop coal from Shields was used, imported through Suine Haven. The salt produced was fine, white and highly prized for curing fish. 'When the maures grow greate the Salt makers remove more easte and come nearer to the Sea and then the former maures become in some few years good pasture groundes.' It was on these maures or holmes that the salters grazed sheep and cattle, the other part of their dual occupation. In the 1590s salter George Brather of North Coates had 160 sheep and 46 cattle. The saltern mounds of industrial waste between Tetney and North Somercotes cover about five square miles and represent some 8-10 million cubic yards of silty sand from which the salt has been removed over a period of perhaps 800-1,000 years. The manufactured salt was distributed by middlemen to inland markets, using the ancient trackway—Salter's Lane—from Beacon Hill at Wragholme over the Wolds through Ludford to Lincoln.

On the Wash shores of Lincolnshire the mediaeval industry was more widespread. The long narrow strips extending seawards from the seadyke or Saltersgate were known as tofts or tufts at Wrangle, Friskney and Wainfleet, with the upper foreshore called Sandacre or greva. The area is still known as The Tofts, and names of original owners survive—Key's Toft and Sigtoft. The salt cote or boiling house had mud walls, timber door and windows and a reed thatched roof secured with rush ropes. The settling ponds for collecting sea water were known locally as 'floors'. Each saltern had sufficient common pasture in the fen to maintain oxen and horses, used particularly for carting away the blocks of salt wrapped in woollen cloth. Boats were also used, especially to bring in peat from the fenland turbaries (peat turf grounds), such as those in the East Fen inland from Wainfleet. Further south, as Bicker Haven and the estuaries of the Welland-Glen and the Nene began to silt up, the saltmakers had to move nearer the sea, and new works, with further extensive mounds of waste were established at Pinchbeck (Hills Marsh), Saracen's Head (at the mouth of the Whaplode river), at Moulton (in Pudingneuland), along Holbeach Bank and Gedney Dyke, and—one of the greatest saltmaking areas—the mouth of the Fleet river.

Many of the salterns were owned by monasteries, perhaps having been given originally—such as the island of Seilholm in Wainfleet given to the Abbey of Bury St Edmund in 1165 by Matthew de Praires, seven salterns in Wrangle with pasture and turbary given to the Abbey of Waltham Holy Cross in the late 12th century by Simon le Bret, and two saltpans in Friskney given to Bardney Abbey in the mid-13th century by Philip de Kyme. In general it was a cottager's industry, some making a living from saltmaking alone, although there were some richer peasants, particularly in Fleet, where one of the leading families in 1315 were the Neels, with 88 acres of land and 13 salterns. The annual rents paid in salt give some idea of the volume of production—200 bushels in Pinchbeck in 1274 and 160 bushels in Fleet in 1316. Usually it was paid in two instalments: between 1175 and 1193 Thomas de Fleet paid Thorney Abbey (they held the greatest interest in that shoreline) 12 middes (bushels) of salt by the Spalding measure to the cellarer, half at Lady Day and half at Michaelmas, and in 1321 Laurence de Holbeach paid one rose and 80 quarters of salt at 18d a quarter at the Feasts of St Mary Magdalene (22 July) and Nativity of the Blessed Virgin Mary (8 September); between 1240 and 1250 Simon de Friskney paid Bardney Abbey seven bushels of salt, 9d, two hens and a cock, the salt and money at the Feasts of St Botolph (17 June) and St Oswald (5 August) and the poultry at Christmas.

Saltmaking on the Wash shores depended on the supply of large quantities of turves from the fen. To be fully productive the turbaries needed to be neither too wet nor too dry. This balance was disturbed in the 13th century, which saw a combination of wet seasons and a rise of sea level which caused flooding of the turbaries. The saltworks themselves were affected by abnormally high tides: in 1253 'about the feast of saint Dionisius a great flood took place in Holland'; on 1 January 1287 'nearly the whole district (of the Fens) was converted into a lake' followed by sea floods on 15 December, and 4 February 1288. One effect of this stormy period was the creation of a storm beach and dunes on the north-east coast of Somercotes (Warren), but it was possible to re-establish salterns on the new shoreline.

Although the Lincolnshire saltmaking industry survived those setbacks, it was less productive. The great sea-flood of 5 October 1571 when 'all Saltcotes, where the chief and finest salt was made, were utterly destroyed' did not help an industry which was already suffering from the competition of cheaper production elsewhere. Poorer quality Bay salt had been imported from the Bay of Bourgneuf on the west coast of France from the 13th century. But it was the production of salt from the Tyne and Wear, where plentiful coal was used to boil the brine, and particularly from the Firth of Forth because there was no tax on Scottish salt after James I came to England, which was the deciding factor. The salters of Donington and Wigtoft were able to continue longer than most, by importing Cheshire salt from Droitwich brine wells, by pack-horse train along the Salters' Way, crossing the Lincoln Heath from Saltersford at Grantham and the Holland Fen by the Bridge End Causeway. The salt was refined with sea water to produce a very fine white product.

Another equally shortlived method of producing salt was tried in the late 16th and early 17th centuries at Sutton, Chapel and Ingoldmells. This was the Sunworks, the remains of which are to be seen when storms have scoured sand from the clay foreshore, although most of the rectangular excavations are the result of digging clay to repair post-mediaeval sea banks. A sluice in the protective sea bank allowed a stream of water to flow up a ditch and into large shallow pans, 60 feet wide and about 100 feet long, which had been cut into the clay. These were sealed off and the water allowed to evaporate naturally for a time. The sea water was then slowly drained into a smaller pan at a slightly lower level where evaporation again occurred, and then into a still smaller and lower pan. By this time, given good weather, the brine was strong enough to be boiled, and the salt hung in cone-shaped baskets on the wall of the boiling house to drain and dry. But by 1640 the Lincolnshire salt industry was finally dead.

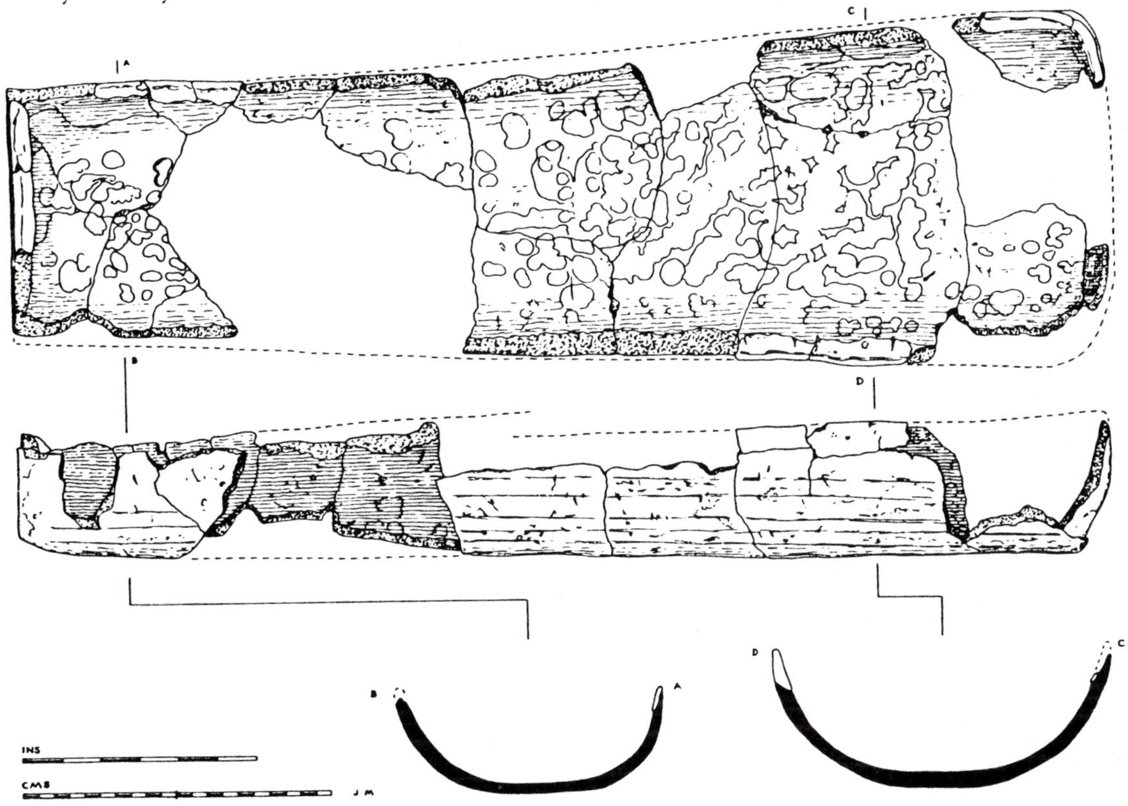

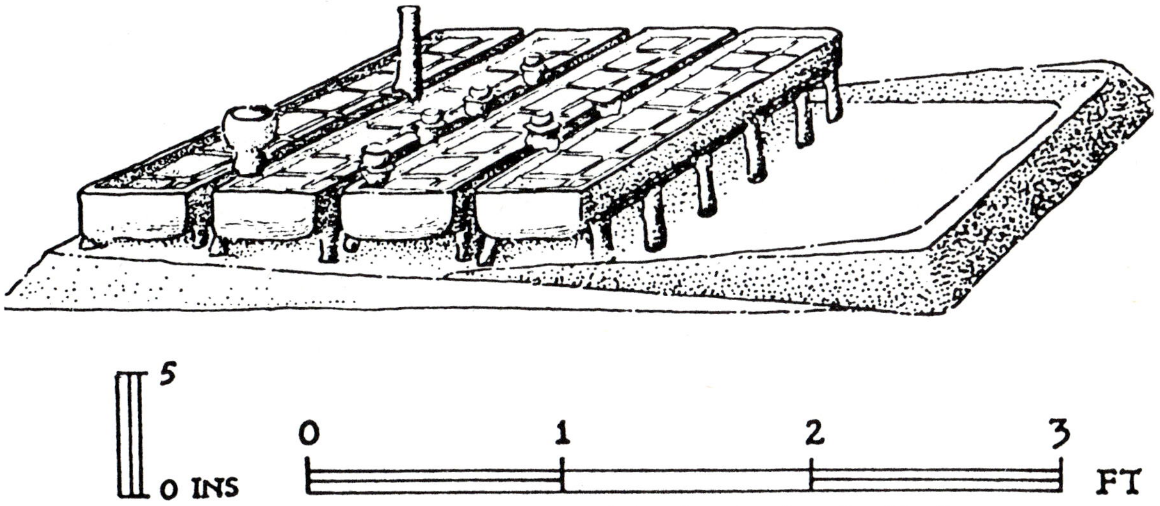

ABOVE: Pottery salt evaporating trough from Iron Age site at Ingoldmells. (JM) BELOW: Reconstruction of Roman salt-making hearth, Ingoldmells. (HLC)

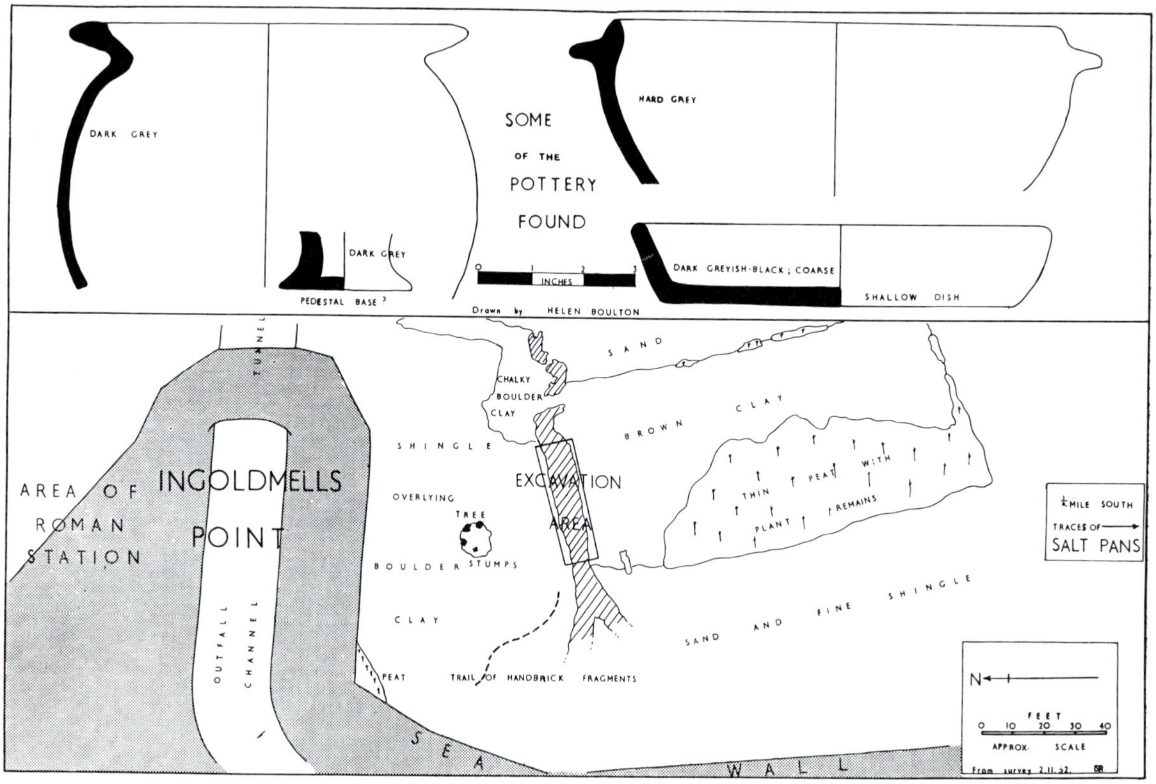

LEFT: Excavation of a Romano-British salt making site on the beach at Ingoldmells, 1932. RIGHT: Grassed over saltern mounds at Grainthorpe. (EHR) BELOW: Site of excavation of Romano-British site near Ingoldmells Point 1952, with some pottery found.

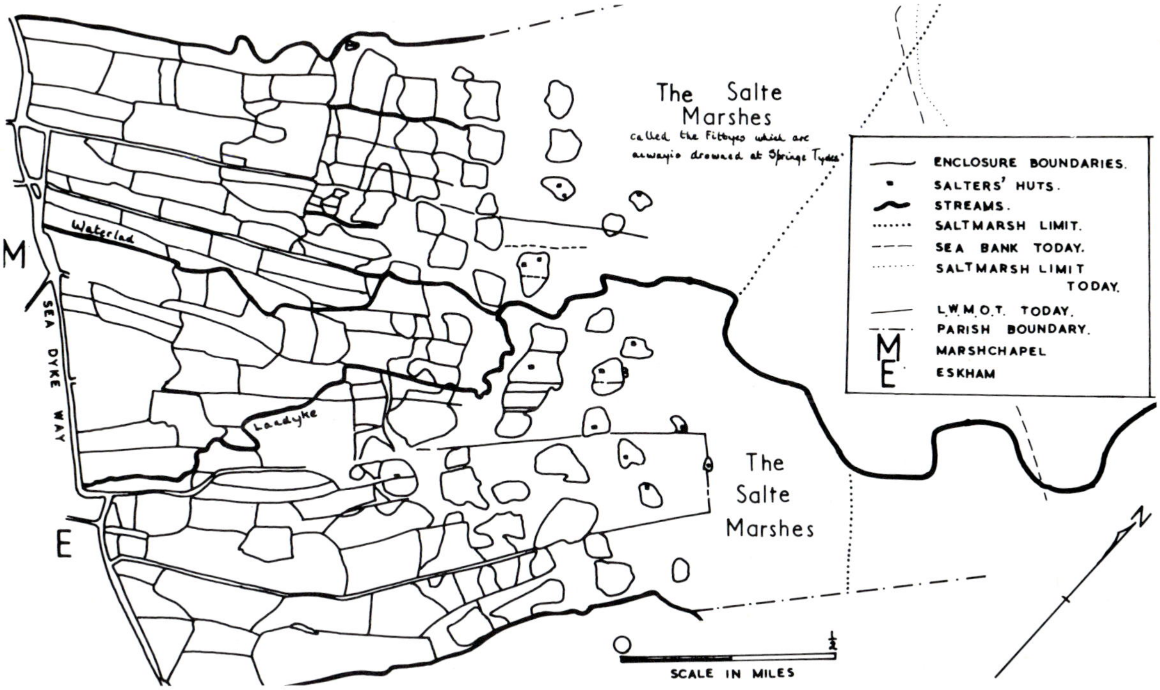

ABOVE: William Haiiwarde's map of the saltworkings at Marshchapel, 1595. BELOW: Reconstruction of a mediaeval saltcote at Marshchapel.

ABOVE: Sunpans and building sites exposed on the shore at Sutton on Sea after the 1953 storm-flood. BELOW: A crust of salt on the shore when dry weather followed high spring tides.

The Smugglers

The night was dark but not stormy. Jack Anderson, the new farm labourer hired by Ben Oliver at Louth Stattus Fair, and Ned, who knew what it was all about, were ordered to get the horses and carts ready. It was a short distance from the farm to Oliver's Gap north of Mablethorpe, the soft sand deadening the sound of wheels, and then across the wide beach towards the sea. When the gently lapping water was up to the bellies of the horses and the body of the cart they turned the carts round with the horses' heads towards land, and waited. Presently a small Dutch galliot sailed quietly alongside. Kegs and bales were transferred until the cart was full, and the vessel slid away to the next cart. As the carts lumbered slowly but silently over the gap in the sandhills, the carts of other neighbours passed by almost unseen in the moonless night. The organisation had been good and there was no Riding Officer of the Preventive Service about. It was a successful night for smugglers.

A few nights later the horses were again harnessed, the white face of one covered with sacking, and the cart wheels wrapped in straw, and some old sacks containing sea-sand were put in the cart just in case the straw wore off the wheels. They had to traverse some rough chalk and gravel roads across the Marsh and the noise of iron-shod wheels could carry some distance and arouse suspicions. Before morning the gin and tobacco would be safely stowed away in cottage chimneys, barns and even rectory cellars.

The hey-day of smuggling, or contraband trading, was in the 18th and first half of the 19th centuries. The vast increase in customs duty on tea, spirits, tobacco and silk made it profitable to 'import' them without paying duty. A tax could also be levied on exports, or penalties imposed for illegal export. When there was an embargo on the export of wool in 1274-75, William de Len and certain other Louth men sold 200 sacks of wool to Flemish merchants and got them away by ship from Saltfleet and Suine with the aid of bribes, including a cask of wine for Alan of Conisholme. Later, under an act of 1672, the export of wool was prohibited; increasing amounts of wool unsold in the county and the demand from the Continent led to smuggling; the Lincolnshire coast, with its gently shelving sands and secluded creeks, was well suited for the purpose. The wool smugglers, who only worked at night and were known as 'owlers', operated from marshland villages near the coast like Irby in the Marsh near Wainfleet, and small boats put out from Grainthorpe and Saltfleet to meet ships out at sea. It was said that few ships left a Lincolnshire port without one or two sacks of wool, perhaps labelled 'hops', surreptitiously stowed in the hold.

In 1785 no less than 90 tons of wool were put aboard a vessel at Goxhill Haven in the Humber estuary, for export to Dunkirk. That same year the first Stuff Ball was held, at Alford, as an attempt to stimulate interest in the manufacture of wool in the county, and two years earlier the philanthropic rector of Theddlethorpe and Willoughby, Rev Reynold Gideon Bouyer, started local spinning schools. But these and other measures did little to check the smugglers. And what was more natural than to organise a return cargo? The smuggling-out of wool suffered a decline when the already harsh penalties for illegal export, including flogging, transportation and hanging, were increased in the late 1780s, and the efficiency of the preventive services improved. But the smuggling of imports was already well established, and continued after prohibition of wool export was removed in 1825. In fact, so much Dutch gin or hollands was smuggled into Dent's Creek on the Humber, in kegs and bladders sometimes weighted and hidden in the creek itself, that it became known as New Holland.

It was the task of the Customs and Excise to prevent evasion of duty payment by smuggling. The Riding Officers were the oldest branch of the preventive service; formed in 1698, their function was to prevent smuggled goods, which had eluded seagoing Revenue cruisers and Customs officers ashore, from being moved inland. They patrolled up to ten miles inland to seek hiding places. Perhaps their inefficiency was a consequence of the bribery and corruption in public life at the time. Thomas Paine, author of *Rights of Man,* was an exciseman set to watch smugglers at Alford in the 1760s and dismissed for misconduct. Even when a smuggler was caught and brought before the local magistrates, he was likely to be dealt with leniently; they did not wish to put in jeopardy their own supplies of Congo tea, brandy and cigars. During the Napoleonic wars the Preventive Water Guard was formed (1809), but even then there were only nine cruisers and thirteen boats for the entire east coast of England. One 'open boat in the pay of the Government at Boston is the height of burlesque' claimed Edward Walls of Spilsby in 1795. He feared smuggling piratical vessels would land for pillage and plunder, and thought it not unreasonable that a cutter carrying eight or ten guns should cruise continually between Wainfleet and Saltfleet Haven.

After the wars smuggling into Lincolnshire was largely of tobacco. Cigars were rather fragile for the rough and tumble of smuggling, although a large quantity was found concealed in a stack on a farm near Boston in 1832. Seven years later 600lbs of pipe tobacco captured by the Excise was auctioned at Louth, the whole lot bought by grocer Isaac Smith. Unmanufactured tobacco, on the other hand required less careful handling, was more profitable and could be readily disposed of to manufactories. It was packed in a hogshead holding 1,300lb but this was usually broken down at the smuggling port in north-west Europe into bales of 50-100lbs for easier handling. The boats used in smuggling were made in those same ports. The 'fastest and finest boats ever built', and ideal for smuggling in an era when speed replaced armament as the defence against revenue cutters, were the luggers—often three-masted lug-rigged fishing smacks.

In 1828 the smuggling lugger *Maria* of Flushing was finally captured off the Lincolnshire coast after a long career. In November 1833 'the Customs House officer stationed at Ingoldmells, in company with Mr James Smith of Wainfleet, agent to Lloyd's, captured a fine new smuggling vessel of sixty tons burthen. The greater part of the valuable cargo had previously been landed and despatched beyond the scrutinising eye of the officer. The crew being strangers to the coast had run the vessel ashore, and she would soon have become a total wreck. Four Frenchmen were taken to a Public House near the spot, on suspicion of belonging to the vessel, but they will most probably be discharged from custody in default of proof'. Ingoldmells was in fact a favourite smugglers' landfall into the second half of the 19th century. Perhaps because in Leila's Cottage lived James Waite, a notorious smuggler who not only suffered jail but three of his boats were sawn in half in attempts to curtail his activities. And not far away had been Joseph Lowe who kept the Scabbed Lamb Inn at Jackson's Corner; in 1834 he was fined the then enormous sum of £1,000 for smuggling.

The Preventive Service was reorganised into the Coast Guard in 1822. Two years later Lieut Butcher and the revenue-cutter *Redbreast* made two notable captures near Wainfleet Haven. In February a cargo comprising 165 tubs of Geneva gin, 29 bales of tobacco, two boxes of 'Segars' and a box of playing cards (duty payable 2s 6d a pack) was taken; three of the smuggling vessel's crew were taken prisoner and one escaped. Later in the year the cargo seized included Geneva gin, tobacco, snuff, tea, sugar, pictures and china; three of the crew were committed to Lincoln Castle. Another 'who says he was employed by a horse-dealer to go over to Holland with some horses from Horncastle fair, and was only a passenger on the way home', was remanded until he could prove it. Smugglers running into Sutton Wash in the frosty spring of 1828 got into difficulties of a different sort. Leaving the lugger at anchor offshore, a 30-foot boat heavily laden with Holland gin and tobacco became stranded in the ice and the crew had to abandon boat and cargo. Furbishing their rusty cutlasses and mustering their courage, the local Customs officers took possession of the boat and employed carts to unload the contents. As one of the carts was being hauled up the steep and mirey slipway by the Ship and Cross Keys Inn, someone pulled out the tip-stick. Up went the cart, scattering the contents in the mud and in the ensuing scramble several casks of liquor were 'run away with'.

In 1842 the revenue cruisers were replaced by a small force stationed at Skegness, Saltfleet and Tetney Lock (at the entrance to the Louth Navigation Canal). In August 1846 the Dutch smack *Unterehmung* attempted to land nine bales of unmanufactured tobacco from a small boat to men waiting with horse and cart on the beach at Ingoldmells. The Revenue cutter put out to sea but the smugglers threw the bales overboard. These were washed up on the shore and taken by waggon under escort to the Customs House at Boston. When the smack was seized, 22 more bales (totalling 5000lbs) were found, divided into convenient packages for carrying.

The creeks and marshes of the Wash and the Humber estuary may have had their share of the 'coastal trade', but most successful smuggling was accomplished on the open sandy shores between the Customs stations of Saltfleet and Skegness. Here the shore was backed by sandhills with gaps over which the carts could be pulled to and from the beach—known as 'pullovers'. Along this stretch of coast the locals were the most skilled and resourceful in avoiding detection. While still in his teens in the late 1840s, young George of Sutton in the Marsh first went with a party of local smugglers to Holland to fetch a cargo of tobacco (unmanufactured at 6s a lb and shag at 7d). Although this was thrown overboard when they feared detection, they later made successful runs at Huttoft and Mumby Chapel. In August 1850 they were caught in a gale on the way back from Holland; the third night the wind dropped and they made landfall at Skegness, but it became so misty they could not see the signal. Retreating, their boat ran near a revenue cutter in the mist, so they slipped up Boston Deeps. Unfortunately the boat was spotted by the Wainfleet coastguard boat which fired a gun for them to lay to. They had just enough time to dump the tobacco overboard.

A common ruse to decoy the coastguard away from the actual landing, which worked more often than not, was to draw his attention to a mysterious boat showing a light and sailing coast-wise in the opposite direction. Or a lantern would be strapped to a horse's leg 'so that coastguards might mistake the animal for a ship riding at anchor'. On one occasion smuggler Tommy feigned to be in his cups in a pub in North Somercotes, and rambled on to the effect that if only they knew it, the coastguard of the district could capture a prize contraband the evening of that very day. It worked. The word was passed on and preventive officers from the next station at East Theddlethorpe came to reinforce the vigil. That night Tommy supervised a successful run on the unguarded coast. A rather neat trick was when a coastguard officer was 'treated' by known smugglers at a wayside Tom and Jerry (inn) and he began to boast that he could run down anyone even if it meant jumping over the gout (drain by the sluice). On being challenged to demonstrate his prowess, inevitably he failed. While he went home to change into dry clothes, another successful run was made. And of course there were some preventive officers not above taking a bribe of smuggled spirits as the price of turning a blind eye; in fact the old saying is still quoted—'Put a guinea in each of my eyes, and I shall see nothing'. Still, could it be that the skeleton with royal insignia brass buttons, found in 1902 in the brickwork of the Vine at Skegness, was that of a 'preventive officer' who refused such bribery?

It was often the case that all the contraband could not be moved off the beach in one night, and some would be hidden in prepared sand caves and other secluded spots in the dunes. It could be removed in daylight, by going to the shore with a barrow, on the pretext of taking sand and shingle for road repairs. Some, hidden in a hurry, they never did return for; at the beginning of this century the recognisable remains of a hogshead of tobacco were found in the dunes near Mablethorpe pullover. Even bathing vans at the infant holiday resorts were used as temporary hiding places. More usually the contraband was hidden in chimney recesses, roof cavities, cellars and hollowed-out haystacks, or even hung from hooks on the underside of bridges over drains. Two farm houses near Oliver's Gap, Bleak House and North End Farm, the homes of smugglers Ned Bell and William Twigg, had their secret places.

One night the farm waggon was being used to move contraband inland from Oliver's Gap, taking the usual precaution of having a man following, out of sight in the fields. It was just as well, because the road was being watched. Out jumped a 'Preventive' who ordered the driver to stop, but he pushed on, trusting to the darkness for cover. 'Stop or I shoot!' cried the officer, but still there was no response. He fired and the driver fell to the ground. The officer promptly rushed off to report and get assistance. When he returned with the Coastguard, cart, horses and driver were nowhere to be found. The shot had in fact missed.

About a mile north of Oliver's Gap was (and still is) the Curlew, a brick cottage in the dunes, ideally situated for smuggling, with a clear view to the sea in one direction, and out over the Marsh in the other, where a preventive officer could easily be spotted. Coastguard Gallakin nearly surprised the household, and it was only the quick-thinking daughter who kept him talking (he was sweet on her anyway), to gain time for the men in the sand cave behind the cottage. At another farm the family were having dinner when the Riding Officer called, suspecting contraband was in the house. He was invited to 'have a drop'; in fact he had a drop too many and fell asleep, thus allowing the sons to remove the smuggled goods from the bedrooms.

One method of getting tobacco to Lincoln, particularly after the manufactory there opened in 1842, was to use carriers, particularly those of Louth and Horncastle. The transfer was made in some secluded back lane, and the packages disguised by false labels. In January 1845 several Horncastle carriers were caught in possession of nearly three tons of contraband 'in packages representing various articles of merchandise and furniture'. Perhaps someone had informed—like the previous September when 730lb of tobacco, having been landed at Huttoft and transported through Horncastle, was seized as it entered the yard of the Lincoln manufactory. The alternative was to use waggons belonging to smuggling farmers and tradesmen, travelling by back lanes and using staging depots to avoid moving in daylight. Routes included one from Mumby Chapel along the Burlands Beck to Claxby, with an underground brick chamber under a potting shed in the grounds of the Hall, and from Theddlethorpe to South Thoresby and on to Lincoln over the Wolds. Transit depots were in a hen roost at Vicarage Farm at Stewton near Louth, at Scamblesby, at a lone house between Ashby and Horncastle (probably Milestone House) and the Midge Inn at Hatton between Horncastle and Wragby. The rector of Gayton le Marsh went one better—he carried contraband from the ship in his pony and trap and stored it in the 'steel' room in the rectory.

Mary West, then a little girl, lived in a lonely cottage in the sandhills at Mablethorpe. One night her father was out on a 'run' when there was a fight between the smuggling smack and a revenue cutter. Shots were fired and a wounded seaman was brought back to the cottage. Next day the old doctor arrived. He was discreet, asked no questions, treated the wound, and next morning found a keg of spirits on his doorstep. A few days later Mary was perched on a cartload of potatoes as her father led the horse to an inn at Alford, where the cart was promptly backed into the coach-house and the door locked. Under the potatoes were kegs of brandy and under the coach-house was a vault. Mary had grown up by the time Jack Anderson came to work for Ben Oliver. In due time they married. She outlived him by many years and was still in her parents' cottage when all that was heard of smuggling were tales of the past.

'. . . a small Dutch galiot sailed quietly alongside . . . kegs and bales were transferred . . .'

ABOVE: North End Farm, Mablethorpe—the home of smuggler William Twigg. BELOW: Curlew Cottage, Theddlethorpe. OPPOSITE: Mitchell's sailing chart for the Lincolnshire coast 1765.

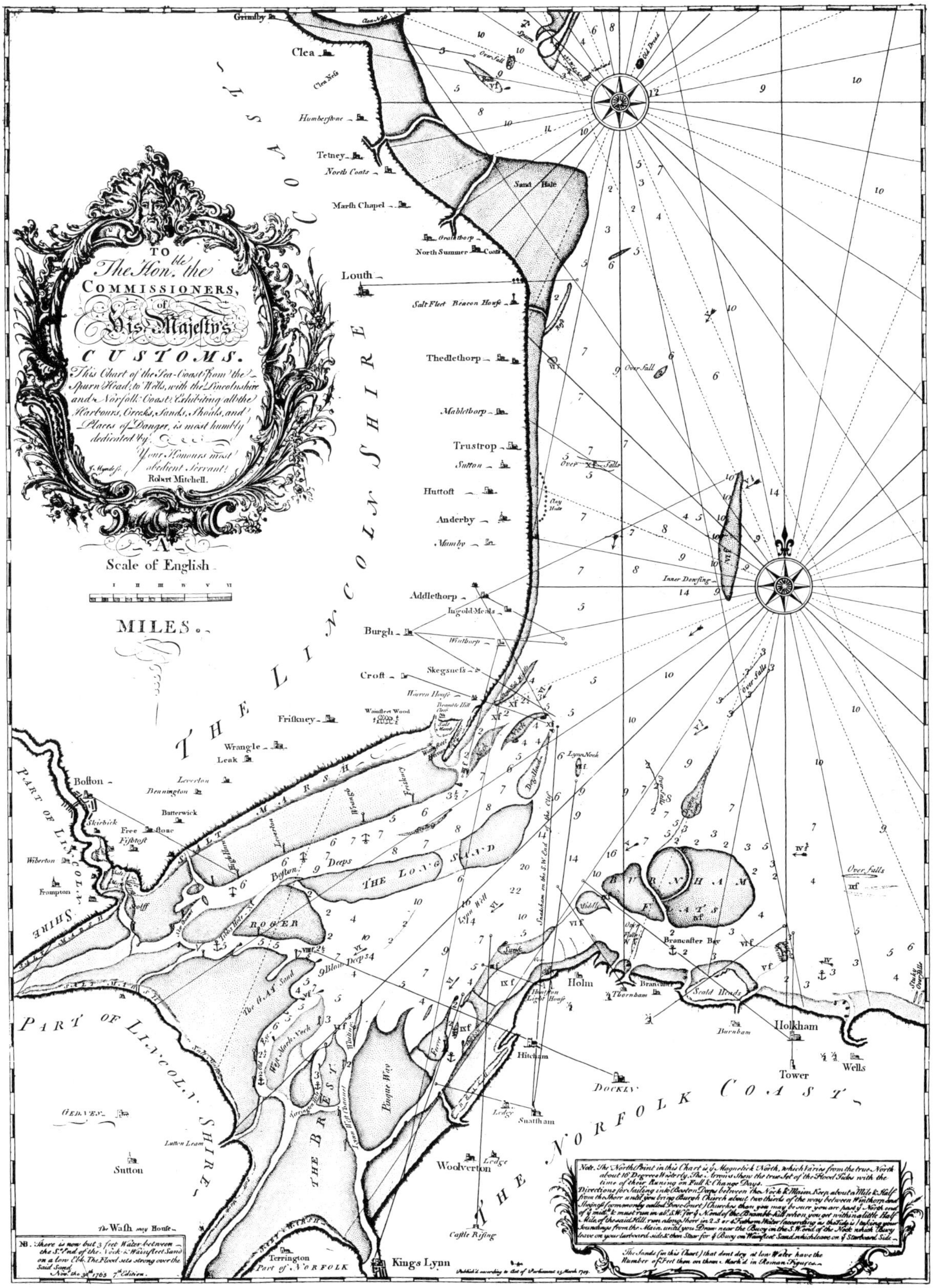
TO
The Hon.ble the
COMMISSIONERS,
of
His Majesty's
CUSTOMS.
This Chart of the Sea-Coast from the Spurn Head; to Wells, with the Lincolnshire and Norfolk Coast; Exhibiting all the Harbours, Creeks, Sands, Shoals, and Places of Danger, is most humbly dedicated by
Your Honours most obedient Servant
Robert Mitchell.
A Scale of English MILES.
THE LINCOLN SHIRE COAST
THE NORFOLK COAST
PART OF LINCOLN SHIRE
Part of NORFOLK
THE LONG SAND
BURNHAM FLATS
THE BREST
Grimsby
Clea
Humberstone
Tetney
North Coats
Marsh Chapel
North Summer Coats
Louth
Salt Fleet Beacon House
Thedlethorp
Mabletborp
Trustrop
Sutton
Huttoft
Anderby
Addlethorp
Ingold Meals
Burgh
Croft
Skegsness
Friskney
Wrangle
Leak
Boston
Benington
Butterwick
Free stone
Fishtoft
Wiberton
Frampton
Sutton
Kings Lynn
Terrington
Castle Rising
Woolverton
Snattham
Hitcham
Holm
Thornham
Brancaster
Brancaster Bay
Holkham
Burnham
Wells
Tower
Inner Dowsing
Over Falls
Boston Deeps
The Wash way House
NB. There is now but 3 feet Water between the S.o End of the Nock & Wainfleet Sand on a low Ebb. The Flood sets strong over the Said Sand.
Nov.r the 30th 1763 7th Edition.
Note. The North Point in this Chart is ye Magnetick North, which Varies from the true North about 16 Degrees Westerly. The Arrows shew the true Set of the Flood Tides with the time of their Runing on Full & Change Days.
Directions for sailing into Boston Deeps between the Nock & Main. Keep about a Mile & Half from the Shore until you bring Burgh Church about two thirds of the way between Winthorp and Skegness (commonly called Dove Court) Churches then you may be sure you are past ye North end of ye midd.le & must run in ab.t S.W. for ye N. end of the Bramble-hill when you get within a little Half Mile of the said Hill, run along Shore in 2.3 or 4 Fathom Water (according as the Tide is) taking your Soundings from the Main, until you Draw near the Buoy on the S.W. end of the Nock which Buoy leave on your larboard side & then Steer for ye Buoy on Wainfleet Sand, which leave on ye Starboard Side.
The Sands (in this Chart) that dont dry at low Water have the Number of Feet then on them Mark'd in Roman Figures.

LEFT: The Anchor Inn or Plummers Hotel, Freiston Shore. RIGHT: Wooden bridge from the long room at the Plummers to the sea bank. CENTRE: Horse races on the beach at Freiston Shore on 29 June 1844. BELOW LEFT: The sea still came up to the bank at Freiston Shore in 1908: Plummers Hotel (left) and Marine Hotel. RIGHT; 'The Good Little Boy'—*Punch* cartoon by John Leech, 1850.

Neptune's Ale

After drinking so much of Neptune's Ale, we are as hungry as hawks . . . and drink like fishes'.

Richard Fowke (July 1805)

Richard Fowke was a tenant farmer in Leicestershire. He had decided to spend a month at Freeston (Freiston) Shore, 'the Brighton of the middle classes in Lincolnshire'—an excessively lavish description of the two hotels (really inns) behind a clay sea wall and looking out over vast expanses of sand and samphire towards the Wash. It took him three days by coach and chaise to reach the Coach & Horses inn (later the Anchor and now reverted to the family name associated with it in the 19th century—Plummers). Board was 5s a day. 'Many of the bathers spent 10s 6d, but what is that to a wine bibber?' Sea-bathing there was reputed to be good for 'all scorbutic complaints, sore legs, sore eyes, surfeits, hard drinking, nervous habits, hydrophobia &c, and a poor appetite'. Yet Fowke thought 'a good many of the company came to see and be seen more than for sea bathing. Young ladies to see for a husband, and young fortune hunters for wives'. He had 'a very neat lodging room on the first floor' with a view of 'the phenomenon of the moonshine on the tide', and 'the advance of the waves clad in silver gray'.

Also on the first floor was the dining room where he ate 'twice a day of sole, and shrimps by the thousand.' Over the stables was a long room where the Sunday excursionists from Boston drank tea, with wooden bridges giving access to the sea bank for the promenade. Or there were two bathing machines (changing rooms on wheels) and the sand for riding. When Fowke left for home, there were 32 guests—mainly from Leicestershire, Northants, Rutland and Nottinghamshire.

The drinking of seawater, or Neptune's Ale—for medicinal purposes—was first recommended in 1667 by Dr Robert Wittie of Scarborough Spa. He said 'it cured gout and killed all manner of worms'. Nearly a century later, in 1752, Dr Richard Russell produced his 'Dissertation on the Use of Sea-Water'. This prescribed 'a pill compounded of some or all of crabs' eyes, burnt sponge, vipers' flesh, cuttle-fish bones, snails, tar and tincture of woodlice taken at night with a pint of sea water in the morning'. Even Lincolnshire's Rev John Wesley in his *Primitive Physic* (1776) recommended taking sea water every other day for swelled glands in the neck, and for shingles 'drink sea water every morning for a week; towards the close, bathe also'.

For the dip in the sea not only was 'a perfect repose of the Body and calmness of mind' required that the pores of the skin should be closed. Thus the colder the weather the better, and 7am in November was recommended. And it should be remembered that it was nude bathing. According to Edward Baynard in *Neptune the Physician*.

'Cold bathing has this good alone:
It makes old John to hug old Joan!
And does fresh kindnesses entail
On a wife tasteless, old and stale.'

By the mid-18th century not only was the bathing machine (with or without a modesty hood) essential equipment for the seaside inn, but ladies could hire long concealing garments to wear in the water:

'The ladies dressed in flannel cases
Show nothing but their handsome faces.'

Despite the fact that summer was becoming the more acceptable season, the services of a dipper were usually required to duck the reluctant bather. When W. M. Childs of Alvingham near Louth was a child in the 1870s, he was taken with others in decorated farm waggons to Mablethorpe. 'Some of us bathed,' he recalled, 'but the experience was disenchanting. We were boxed up in dank and dark machines, dragged by a horse an alarming distance out to sea and were there seized by a brawny bathing-woman, robed in blue serge, and forcibly ducked.'

The cult of sea bathing came relatively late to Lincolnshire because of the remoteness of the coast and the lack of decent roads. The advent of the turnpikes helped, and the popularity of Freiston Shore required a short extension of the turnpike from Boston. The Ship Inn at Fosdyke Wash, where the turnpike crosses the sandy Welland estuary, was well placed; by 1790 it was a well-patronised bathing and tea-drinking resort for Sunday afternoon excursionists and not uncommon to see 'above a thousand at a time, who bathe, drink tea &c, and return the same evening'. The construction of training walls for the Welland in the late 1830s and consequent rapid growth of saltmarsh removed its bathing potential. At Skirbeck, a mile or so down the Witham estuary from Boston, there were two sea bathing establishments. In 1816 ,S.Newton of Pudding-pie House advertised improvements to his 'Bathing Houses, by which means ladies and gentlemen can bathe at any hour of the day'. At the Hot & Cold Bathing House, Neptune Inn, Isaac Frith offered board and lodging at a guinea a week, children under 12 years half price, and bathing if ladies and gentlemen find their own dresses 3d, or if found by the proprietor 6d. Or you could bathe from Robert Burton's house at Boston Scalp at the mouth of the estuary.

Sea bathing at Freiston Shore lasted for something over a century. The other hotel (both appear to have been built just after the middle of the 18th century) was the Marine, or Nottingham House. In 1845 proprietor John Baker advertised 'Hot & Cold Baths and Bathing Machines always in readiness. Liquors of the best quality, and Wines of Superior vintage. Stabling for upwards of 100 horses, Private Boxes, and Lock-up Coach-houses'. As an added inducement three years later the new proprietor, C. Kinnersley, offered to meet the packet boat from Lincoln at Boston, with free transport to Freiston Shore. The Nottingham connection is interesting. In 1821 Nottingham coachmaster T. Russell inaugurated a thrice-weekly coach service, *The British Queen* from the Black Boy to Freiston Shore. The following year J. Westland at the Anchor Inn reduced his prices 'in consideration of the depression of the times'. Freiston Shore also boasted a number of comfortable lodging houses (W. Gunniss had hot and cold baths in his premises) and a salubrious air. From the 1840s to 1870s there was an annual Sand Fair in June or July, with horse and pony races, amusements and a quadrille band. It was never more than a local resort for Boston after the arrival of the railway at Skegness in 1873, but all came to a rather abrupt halt with the opening of the Witham Cut in 1885 and consequent spread of saltmarsh. Today the Marine Hotel is a forlorn shell and the saltmarsh where the Sand Fair was held has been reclaimed, but the Plummers survives.

Whether or not there were sea drinkers at the Skegness Hotel (later Enderby's—after T. & M. Enderby 1824-50—and then, as now, the Vine) there were races on the sands opposite from at least as early as 1829. The oldest part of the building is the T-shaped section at the north end which could be of the 17th century, since when it has had many additions. Joseph Dickinson was advertising accommodation for 'gentlemen, ladies, and others' and 'as clean a shore as any in England' in 1772. Bathing machines were there in 1784 when Samuel and Ann Salter were proprietors, but when the Hon John Byng arrived in summer 1791 he found it a 'vile, shabby bathing place' with 'no garden, no walk, no billiard room; nor anything for comfort, or temptation'. Ten years later George Pigot had made a garden and bowling green. The hotel had a large dining room, drawing room, several tea and coffee rooms and over 20 bedrooms. The shore was 'bold and firm for Bathing, Fishing and Exercise, and allowed to be the best situation for Sea Bathing in the County'. For the 1805 season John Stafford had installed 'improved convenience for heating Sea Water for Bathing'.

By August 1806 the Skegness Hotel had clearly arrived on the social scene. 'A company more than usually numerous and respectable this year honour the Hotel at this place with their presence', reported the newspaper, 'and afford evidence of the sanative influence of the air and sea bathing on the Lincolnshire coast. The house has for some time been crowded with resident guests, in addition to whom the succession of arrivals of the day and visitors *en passant* is uninterrupted.' The reporter brought the social scene into focus: 'Lady Ingleby Amcotts presides and with a condescending attention promotes the conviviality and entertainment of the company'. One evening the company was 'amused with the dramatic performance of a deserving family from Wainfleet'. The Dowager Countess of Rothes and the family from Gunby Park dined on another evening, and the following day 'a great influx of visitors from Spilsby and the neighbouring towns . . . augmented the party at dinner to upwards of sixty'.

By this time there was a rival establishment—the New Hotel (built originally, it is said, by Rev Edward Walls as a hostel for his friends) at the eastern end of the old sea bank which is now Skegness High Street. John Petty had acquired a new bathing machine for the hotel in 1805, Thomas Melson added a second when he took over in 1809 and a 'most commodious warm sea-bath,' together with a Shower Bath by 1818. Joseph Hildred took over in 1828, followed by his widow Sarah in 1848 (she added an upstairs billiard room in 1850, and enlarged the hotel in 1855), and her son Charles in 1874, by which time it had acquired the family name.

At the beginning of the 19th century the Lincolnshire seaside was only for the relatively well-to-do, although some of the 23 houses in Skegness took in lodgers. At Winthorpe, a mile from the shore, one house was let ready furnished from the 1780s 'for the reception of a genteel family . . . for the conveniency of sea bathing'. John Stafford made up the bridle road from Croft Bank to his hotel into a 'commodious carriage road' in 1816, and also let the four-bedroomed Signal Post Cottage on the nearby dunes from 1820. This and the hotel were the only buildings between the village and Gibraltar House. Skegness in 1827 was regarded as 'a retired situation, well adapted for the resort of invalids', being less crowded than Freiston. It was 'a place where ease and quiet and comfort may be obtained at a reasonable expense'.

Alfred Tennyson, born at Somersby in 1809, was brought to the Lincolnshire seaside at an early age by his parents. They travelled in a 'booby-hutch' or covered car with the family of Rev Drummond Rawnsley (of Halton Holegate near Spilsby) and stayed at McKinley (or Moat) House. This one-room thick house was built in 1780 (by Edward Walls for his sister) just behind the Green Bank (now Drummond Road) near where is now the Arcadia. It faced the open shore, with a removeable plank bridge from the front door to the bank. 'You ran straight out barefoot on to the sandy bank and so across the beautiful hard sands and through the salt-water creeks to the sea. This at high water was close at hand with tumbling waves and seething waters, but at low tide, far as the eye could reach was nothing but sand . . . to the edge of the far-retreating sea.' Is it these memories in *The Last Tournament*?

> '...as the crest of some slow-arching wave,
> Heard in dead night along the table shore,
> Drops flat, and after the great waters break,
> Whitening for half a league, and thin themselves
> Far over sands marbled with moon and cloud,
> From less and less to nothing.'

Or in *Palace of Art?*

'A still salt pool, lock'd in with bars of sand,
Left on the shore; that hears all night
The plunging seas draw backward from the land
Their moon-led waters white.'

In later years Tennyson put up at Enderby's Hotel (The Vine). He could be seen 'a-raavin' and taavin' upon the sand-hills in his shirt sleeves an' all' composing poetry aloud. It is not surprising that the few inhabitants of remote Gibraltar Point (where his brook, as a brimming river, finally reaches the sea) when finding him thus in the early morning hours, 'thowt he wur craazed'.

As a very young boy, Alfred had been taken with his brothers to the tiny village of Mablethorpe sheltering right under the sea-bank. His father found lodgings there cheaper (than Cleethorpes) but difficult to secure. 'Every pot-bellied grocer and dirty linen draper bespeaks his lodgings from year to year.' Having got the refusal of a comfortable house near the sea in 1813, the owner then said that the Brackenburys who came every year had written that they were coming and he could not disoblige them. 'This was very scurvy treatment', wrote Dr Tennyson. 'D – n the lousy race of Brackenbury through all its branches from end to side. There are no lodgings to be had at Sutton, every dirty cottage runs over. As for inns, you cannot be at all decently accommodated in them. They are stinking filthy places, not fit for a pig.' That year they eventually stayed with Mrs Scamblesby, a mile from the sea, for 25s. Later they stayed at Ingoldsby House, on Quebec Road, under 'the long, low line of tussocked dunes' where Alfred stood on the 'sand-built ridge, and thought that it was the spine-bone of the world.'

When verses by Alfred and Charles were published in 1827 *(Poems by Two Brothers)* they took a carriage to Mablethorpe and shouted themselves hoarse on the shore, 'reciting the poems with no other audience than deaf sand and sounding sea waves'. From 1828 to his last visit in 1843, Alfred stayed wih Mrs Wiliman at Marline Villa (at the pullover end of High Street). Other visitors came in considerable numbers from Lincoln and Nottingham. 'But they saays it is a mighty dool plaace,' according to a Mablethorpe fisherman, 'with a deal o' sand, becos there isn't naw band nor pier... but the wind blaws the poor things a bit, an' they weshes their bodies i' the waaves.'

It is not known when the original Inn or Mablethorpe Hotel (later the Book-in-Hand, gutted by fire in 1981) was built, but it was thatched and whitewashed facing the sea alongside a drain when Thomas Frow had to sell it in 1809. He then became its tenant for a couple of years—advertising 'Hot and Cold Sea Bathing'. Known also as the Castle Inn, the name Book-in-Hand appeared in the late 1830s with the acquisition of a figurehead from a wrecked ship. The accommodation included a large dining room, drawing room, smoking room, two bar rooms, ten bedrooms, hot bathroom and stabling for 30 horses. In 1813 Thomas Borman (from Saltfleet) took over until 1820, when he established a private sea-bathing house at Trusthorpe. New houses and cottages were being built with a view to providing summer lodgings (one had an artesian well of 'the chalybeate water') and the next tenant, King Dawson Queenborough, stepped up the advertising with claims of 'the acknowledged superiority of the Mablethorpe shore, conjoined with good roads and almost daily communication with the London mail'. In 1834 the charges were 4s a day including supper, hot baths 2s 6d, shower baths 1s 6d, beds 1s, and an unusual double bathing machine was available.

Thomas Stewart made extensive alterations in 1854 to create the Family Hotel, and organised a six-day coach service to meet trains at Alford (1s 6d) and Louth (2s 6d). The following year R. Sharpley of Great Carlton built two marine residences expressly for the accommodation of families visiting the sea. Thus it was claimed that Mablethorpe 'situated on the German Ocean, has of late greatly improved as a fashionable Watering Place'. Although its population was only 330, in July 1855 it was 'full of visitors to overflowing' with 120 in residence. The village certainly attracted considerable numbers of day visitors, even before the railway arrived. In August 1842 'upwards of 150 Teetotallers and their friends, availing themselves of the splendid weather, engaged the Rechabite band, and went down in 13 carriages to the Mablethorpe shore, to enjoy the salubrious breezes from the German Ocean. The party from preference bivouacked in several social parties on the sea bank. . . . The contemplation of the sublimities of the ocean gave rise to rational conversation on the power, wisdom and goodness of the Deity. . '. In 1871, when the village population was only 424, an astonishing 4,000 'pleasure seekers' descended on 10 August. 'Wagons and other vehicles with gay trappings arrived early. . . . Withern and Strubby Schools were among the companies. . . . the Primitive Methodists of Alford and Louth held a monster camp on the hills.' Could it possibly have been at Mablethorpe that the gross offence occurred on the Lincolnshire coast in 1811 of a man 'bathing and exposing his person'—for which offence he could be prosecuted? Mixed bathing was an offence, and Mablethorpe had a byelaw requiring bathers to wear one-piece bathing apparel reaching to the knee, which was not repealed until 1947.

The original Sutton Hotel, reputedly established in 1691 and already 'a very fashionable bathing place' by 1802, with rooms commanding a view of the sea, extensive stables, a chalybeate spring and 150 acres of rich pasture, was pulled down in 1805 and the materials sold. It must have been an impressive building as the materials included half a million bricks, Wesmorland slates, doric columns, several hundred yards of Yorkshire flagging, marble slabs and chimney pieces, windows glazed with crown glass and staircases with mahogany rails and bannisters. It was replaced by the Jolly Bacchus Hotel with its 'Hot and Cold Baths, at any hour, well-aired Beds, Wines and Liquors of the best quality and first rate stabling'. Other accommodation was also available at Sutton le Marsh, ranging from 'Two neat Furnished Rooms, with the use of the Kitchen, during the Bathing Season, for any respectable Gentleman and Lady, or two Ladies' in the house of a clergyman, to Sea Bank Cottage with six bedrooms and four sitting rooms.

When he left Mablethorpe in 1813, Thomas Frow went to the New Bathing Hotel at Ingoldmells, and inserted this extraordinary advertisement in the *Lincoln, Rutland and Stamford Mercury*: 'THOMAS FROW repectfully solicits the attention of those, who, either for health or pleasure, may have occasion to visit the sea, and begs leave to inform them that the above-mentioned establishment, accessible by excellent roads, replete with every convenience, upon the most prominent point of the coast, elevated close to a fine beach, within a few yards of the low-water-mark, commanding, even from the pillow and the couch, extensively beautiful and generally uninterrupted prospects of the German Ocean, the Norfolk Coast, and the Lincolnshire, woods and wolds, over the intermediate country, richly diversified and highly picturesque—is now open for the reception and accommodation of company in such a manner as, it is hoped, will afford every satisfaction to all persons who, in partaking of marine and rural enjoyments, may be pleased to honour the advertiser with their favours; for which his most sincere gratitude will be for ever felt and acknowledged. The premises contain a very commodious hot salt-water bath, with a stove in the bath-room.'

This sea residence, later called Marine House and finally The White House, contained a 30 ft dining room, breakfast room, four bedrooms and two for servants, and servants' hall. Horses at grass or hay were charged at 1s per night, and a cow could be agisted if required at 3s 6d per week. There was a hard-water pump and a supply of soft water from a slated roof. The building was just behind the sea bank near Vickers Point, and continued to take in holidaymakers until 1911; it had to be taken down after flooding, in the early 1920s.

Before Thomas Borman moved to Mablethorpe in 1813, he had been at the New Inn at Saltfleet for six years. The T-shaped part of the building was the Old Inn of the 17th century which by 1673 had a reputation as 'a place frequented by the Gentry in the Summer season for the eating of fish'—perhaps more appetising than sea-drinking. A century later Mr and Mrs Sewell announced completion of a new building (the additional wing) with 'every convenience for bathing'. Having ridden on the widest sands he ever rode upon from Theddlethorpe in 1791, John Byng found it 'a wretched inn-bathing-house' although it had been recommended for sea bathing for travellers from the midland counties of Nottingham and Leicester. Perhaps the gentry were resorting to nearby North Somercotes where for the 1788 season, private lodgings for a family of 12 to 16 were available, complete with bathing machine, or to J. R. Medley in Saltfleet who offered 'five chambers with six beds (floors all boarded) and a large dining room'. Thomas Borman seems to have improved matters at the New Inn. The first floor dining room could seat 100; its two bay windows gave a 'beautiful and extensive view of the German Ocean' the waters of which were claimed to be 'remarkably fine and in high repute for its beneficial effects'. The claim to a good road from Louth was more realistic—there was a branch of the Louth Turnpike, with special tolls for those going to Saltfleet for sea-bathing. In the early 1820s the dunes round the inn were levelled and walks laid out with trees (which are still there). However, the ocean view and direct access to the shore were lost in 1854 when the saltmarsh frontage was enclosed.

Cleethorpes as a resort for sea bathing developed from the three fishing hamlets of Itterby, Oole and Thrunscoe—the 'thorpes' of Clee. What made it different from other resorts was the fact that it had a cliff or 'gentle eminence at the mouth of the Humber' said to relieve it from the 'generally monotonous character which belongs to the Lincolnshire coast'. Moreover, 'the salubrity of the air, pure and unsophisticated, whether proceeding from the German Ocean. . . or the Wold hills' rendered it 'a desirable summer retreat for the valetudinarian or the invalid.' This was in the 18th century, although to what extent the powers of recuperation also included a sup of or dip in Neptune's Ale is not clear. But by 1791 when John Byng called, it was 'the best of the Lincolnshire bathing shops'.

The original inn was the Cleethorpes Hotel (later the Dolphin) built about 1760, and run by Mrs Ann Clarke in the first decade of the 19th century, but looking rather different from today. There were only two or three lodging houses in 1803; Benjamin Chapman charged 1s 6d for two gentlemen in one bed, and the same for breakfast and tea, 1s a night for the Best Chamber, and 9d per bed for sleeping the boarding guests of Mrs Clarke (who provided her own linen). Visitors stayed from one night to three weeks. They came not only from Lincolnshire (Barton, Brigg, Gainsborough, Grantham, Market Rasen) but from further afield—Rotherham, Derby and Manchester.

By 1818 William Benson at the Hotel had installed a 'commodious Dutch-tiled Warm Bath and a convenient Dressing-room', and three bathing caravans were available. Ten years later one visitor wrote a 'perfectly disinterested eulogism' in which he hoped that 'any prolixity will be excused, the writer having no other object than that of communicating to invalids and summer ramblers' the delights of Cleethorpes. Travel could be by steamboat from Nottingham to Gainsborough and thence by coach, or steamboat to Hull for 2s in the morning and then steamboat to Grimsby in the afternoon in two hours for 1s 6d. Boating parties were frequently made for exploring the lighthouses at Spurn.

By 1842, even before there were any railways in Lincolnshire, visitors came from London. 'The nature of the shore (which is safe during high, and of a superior promenading quality when low water) renders it very adaptable for the bathing of young persons, of whom several hundreds have been brought down by their parents during the holidays.' The same year, under the Enclosure Act, 2½ acres of sea front were set out for public recreation 'on which is a raised gravel walk near the high cliffs, with seats at proper distances for people to rest themselves upon'. However, the taking in of summer visitors was often a sideline to the main livelihood of the village—fishing. During the summer, 17 smacks dredged oysters from the East Coast and the English Channel, cultivated in oyster beds, pits or lays, extending about a mile along the shore near the low water mark. (There were 50 pits by 1897, cultivating 10-11 million oysters but they came to a sudden end in 1903, when it was found that oysters polluted by sewage outfalls had been the cause of typhoid.) Fish caught were turbot (locally called 'brits'), plaice, soles, smelts, bluebacks, eels, skate, whelks, mussels, cockles, crabs and shrimps, cooked fresh every day (except Sunday) and sold to the summer visitors. The first oyster booth on the sands had been erected by Joseph Grant in 1839. The profit from oysters enabled Grant and Edward Moody to purchase old bathing machines from Spurn and equip them with bathing dresses, towels, 'Glass and Coamb', and even ink. During August 1850 they made an average of 6s or 7s a day—Sundays excepted of course.

By the middle of the century Cleethorpes boasted three inns—Dolphin (with George Locking), Leeds Arms (formerly Crown Inn built 1820, then Crown & Anchor, still with George Burgess) and the more recent Cliff House (with Richard Chapman), and 106 lodging houses, 'fit for the reception of visitors in all classes of society', 20 bathing machines and three warm Bath Houses. But it was now the era of the railway, and within a few decades the complexion of the Lincolnshire seaside would be changed beyond recognition.

LEFT: Skegness in 1826. RIGHT: Enderby's Hotel (The Vine) at Skegness, from the lookout on the dunes, in the 1830s. BELOW LEFT: Leaded stained glass panel from a door at the Vine. RIGHT: The Vine Hotel from the air in the 1940s. The oldest part is the T-shaped section on the right.

LEFT: The one room thick Mckinley House at Skegness where Tennyson stayed. RIGHT: Tennyson's seaside lodgings, Mablethorpe. BELOW: High Street, Skegness on the mediaeval sea bank, seen from where is now Lumley Square, 1860. INSET: Willow china model of Tennyson's House, Mablethorpe.

LEFT: Hildred's Hotel, Skegness, c1890. RIGHT: Book in Hand Hotel, Mablethorpe c1900. CENTRE LEFT: Jolly Bacchus Hotel, Sutton with new extension (left) c1905. RIGHT: Pie in Hand, Mablethorpe. Figurehead was from the barque *Rennata* (wrecked in 1878) adapted to carry a pie and used to advertised Gibbins Duffin's eating house and his home-made pies. BELOW LEFT: Seabank Cottage, Sandilands. RIGHT: All that remained of the White House, Ingoldmells before it was finally demolished.

LEFT: New Inn at Saltfleet. RIGHT: Bathing machines on the shore between Grimsby and Cleethorpes, 1862. BELOW: Mablethorpe beach before the invasion of trippers (water colour by Stella Sharpley).

Day of The Tripper

With the railways came the trippers, dubbed by some 'the less salubrious classes'. And they came in increasingly large numbers. The two railways opened to Grimsby in 1848—the Manchester, Sheffield & Lincolnshire and the East Lincolnshire—which left little over two miles to walk to Cleethorpes or make use of the five times daily horse-drawn omnibus service. The railway companies, particularly the MS & L, laid on cheap day excursions. On 19 June 1850 one train alone brought 2,000 people from Nottingham, the Barrow and Gozhill Sunday Schools took their annual trip and there was 'an extraordinary influx of country visitors'; some 5,000 people were 'on the sands, in the pleasure ground, or disporting themselves in the sea' at a time when the normal population was only 840. Cheap trips were advertised from Lincoln at 3s First Class, 1s 6d Covered carriages (children 9d). 'The number of cheap trips and day visitors keeps this place all alive.' By 1852 there were 32 bathing machines, and the donkey on the sands had become an established feature. There were archery butts at Beaconthorpe, the oyster booths (also selling biscuits, cordials and temperance drinks), 'the English Blondin, Clare Davali' who performed on an elevated rope from the highest point of the cliff to poles fixed in the mud. There was Mark Ellis's 'Splendid Bazaar of Fancy Goods', Canty's 'Royal Bazaar—The Very Noah's Ark of a Toy Shop—Selling off daily, at a reduction of 20 per cent', and Clark's 'Victoria Bazaar' with the 'largest and cheapest stock of fancy goods in the United Kingdom'.

There had been some new housing built in Cleethorpes, but sanitation had not kept pace and there was gross overcrowding in the holiday season. In one small house in August 1854 there were 25 people lodging, and in another, three beds were occupied by 14 people. The village was crammed with about 1,300 visitors, when there was a severe epidemic of cholera, and around 80 people died. Conditions were chaotic and the resort was virtually evacuated. Sanitation and drainage were improved and the following year was claimed to be 'exceedingly healthy', the 'pleasure trip trains' started running again, 'goodly companies of ladies and gentlemen are to be seen parading the shore . . . and the whole plant of bathing machines, donkeys and drivers appear to be in full working order'. Cleethorpes was once again 'one of the most eligible and salubrious bathing places in Lincolnshire.'

A single track branch line of the MS & L railway was extended to Cleethorpes in April 1863. The effect was immediate. For Monday 3 August the Primitive Methodists, one of the main driving forces in the commercial development of a village of 1,400 people, organised and advertised their annual tea meeting. An incredible 30,000 people arrived by train: a special packet trip was run from Hull to Grimsby and Oddfellows turned up from all over the Midlands and North. In addition the best part of another 10,000 arrived in some 400 horse-drawn vehicles or walked. Despite the use of many tea tents, catering must have been stretched, to say the least. One of the oyster and refreshment booths was that of Levi Stephenson, one of the original Methodist families of Cleethorpes. In the cholera outbreak he had lost his wife and two children, but he had married again, and bought a boat. The *Cambridge Lass* became a popular feature of the shore—'Come for a sail with old Levi—the man with the big nose'.

The success of railway company promotions such as seaside trips for four or eleven days from Nottingham (5s or 7s 6d return 3rd class) and Sunday (!) cheap trains from Worksop (depart 8.45am) and Rotherham (dep. 7.30am) and return from Cleethorpes 6.30pm necessitated doubling the branch line in 1874. One Saturday in July 1878, 13 special excursion trains from Huddersfield, Nottingham, Lincoln, Manchester, Sheffield and Rotherham brought in 10,000 day trippers; and on a rather dull August Bank Holiday in 1892 there were more than 30,000 people on 21 trains from Burton, Chesterfield, Manchester, Northampton, Parkgate, Penistone, Wigan, Nottingham (3), Leicester (4) and Sheffield (7). The people of Cleethorpes are known as 'Meggies'. It is supposed that this derives from the rail fare from Grimsby to Cleethorpes—a halfpenny, the local slang for which was a 'meg'; regular users would buy a book of 'meggie' tickets.

At Skegness hotel proprietors were quick to point out after 1848 that 'the principal watering place on the Lincolnshire coast' was 'within an easy distance of the Burgh station on the East Lincolnshire Railway', and each had an omnibus to meet trains (fare 1s). Although there could be an influx of people in August, such as 150 children of the Burgh Wesleyan Schools who 'had a glorious day at the sea-side, spent in cricketing, donkey-riding and other appetite-getting fun' (1859) or 3,000 attending the races in 1861, Skegness could still be referred to in 1866 as a 'retired watering place' where 'the village is free from bustle'. The opening of the Wainfleet & Firsby railway in 1871 (26 years after the failure of the Lincoln to Wainfleet Haven proposal) did little to disturb the pattern; Atkin's omnibus met the trains (fare 1s, or 1s 6d return). The Skegness Extension followed two years later, opening on 28 July 1873, just in time to take advantage of the statutory August Bank Holiday under Lubbock's 1871 Act. The Bank Holiday gala of 1874 brought in 10,000 trippers when the population of the village was a little more than 500. Cheap day excursions were run from Nottingham (dep. 7.40am) via Grantham, and half-day specials such as that for the Foresters of Boston (return from Skegness at 9.10pm).

For the Skegness races of 1876 the Great Northern Railway advertised a range of cheap excursions, attracting 10-12,000 trippers. The races were held on the sands in front of Hobson Dunkley's sea View Hotel (built 1862 and enlarged 1873) half a mile north of the main village street. It had accommodation for 1,000 day visitors and sleeping for 200, an 80ft dining room with news, smoking and refreshment rooms, croquet and dancing lawns. A particular feature was a 'promenade or land pier, 160 feet long . . . from the balcony of the hotel to the foreshore.' In May 1875 a charter train from Kirkstead took 220 people for 'boating, bathing and other amusements' at Skegness with a knife and fork tea at the Sea View Hotel. 'The table being cleared, dancing was indulged in to the strains of Messrs Redshaw & Chauntrey's band.' In July 925 children from the Wesleyan Sunday School, Horncastle, spent a day on the shore.

Day excursion traffic built up rapidly: no less than 220,000 excursions went to Skegness in 1878. In June 1880 an excursion train of 38 carriages carried 1,000 scholars of the Blue Coat School, Lincoln with Governors and friends. During the week ending 3 July, over 12,000 passengers arrved when 'the encampment of the South Yorkshire Volunteers was undoubtedly a great and novel attraction'. On 8 July the choirs of St Mary's Church and Trinity Chapel, Horncastle made up a train load of 700 including friends and the band of the Horncastle Volunteers which played on the platform at Boston station as well as during the day at Skegness. Over 9,000 trippers came on the August Bank Holiday; in fact during August there were 11,410 ordinary passengers and 47,068 excursionists. In July 1881 the races were attended by 25,000 people. In July and August 1882 excursionists descended on Skegness at the rate of 10-12,000 a week, with a peak of 20,000 on the August Bank Holiday. At a time when the resident population of Skegness was a mere 1,400, it must have been an astonishing sight. Because the railway was only single track (not doubled until 1900) the last train did not leave until 2.30am on the Tuesday with 'many hundreds spending the night in waiting rooms or in the streets'. The railway completely changed the character of Skegness, not least because 'the thousands of excursionists taken during the season from the manufacturing districts to have a sniff of the ocean leave behind them large sums of money'.

However, the influx of such visitors was looked upon with 'a feeling somewhat akin to dread' by some of the quiet people of Skegness. Occasionally, reported a local newspaper, 'the excursionists drop down upon the place in numbers so overwhelming as to leave the town as bare of nutriment as was Egypt after the visitation of the locusts' (on that 1882 Bank Holiday all the food shops were cleared out by 6pm). Also the habits of some of the excursionists, 'especially the ladies and gentlemen from Sheffield and Oldham' were reckoned to be rather primitive: 'they think nothing of walking into a lady's parlour and informing the astonished resident that they would "loike t'eve some beer"; and both sexes bathe together in hundreds and flounder about in the waves with apparently an utter unconsciousness of impropriety'. It was claimed that the changes in the character of Skegness was 'driving away the respectable families who were wont to visit the quiet spot'. One family which remained loyal included a girl who first spent a holiday in Skegness in 1860 as an 11-year old; as Mrs E. Bradshaw of Hackthorn near Lincoln she was still visiting in 1930, having notched up a remarkable record of 70 successive annual holidays there. Others responded to the advertisements of Vincent Cross Crow of the Bell Hotel at Burgh, who offered an omnibus and carriage service to 'the more fashionable and retired coast of Ingoldmells, Hogsthorpe and Chapel Shores'. Certainly Chapel benefitted, with villas let for the season and the Hotel (the Vine) having increased family custom.

The differences in development of Cleethorpes and Skegness as seaside resorts, and particularly of the seafront amenities, are because one was a coast of erosion and the other of accretion, and because of different land ownership patterns. While speculative builders developed the town of Cleethorpes on small plots allotted at enclosure and on the larger areas owned by Sidney Sussex College, Cambridge (56% of the township) the Urban Sanitary Authority approached the MS & L Railway Co in 1880 (who were enlarging the station) with a plea to preserve the foreshore for recreation, by halting erosion of the cliffs. Work on the sea defences started in 1883 with a wall 6-7 feet above normal spring tide level, thirty yards out from the cliff (despite difficulties caused by patches of running sand), and the area between filled and levelled to give a mile long promenade 65ft wide. The designer was H. B. James and the cost £33,000. A further £10,000 was spent on swimming baths (later the aquarium), refreshment rooms, 3 colonnades of shops, a restaurant and photographic studio; gardens were laid out, including the sham ruin observation point Ross Castle (named after the secretary of the railway company) and the promenade was lit up. For the opening by Prince Albert on 2 July 1885 an arch of Norwegian ice was constructed on Itterby Road (renamed Alexandra Road in honour of the Prince's mother).

Beyond the Brighton Street slipway (at the end of the old Fisherman's Road) the houses along Sea Bank on a low clay cliff were in danger of being washed away. Under the 1902 Cleethorpes Improvement Act four acres of foreshore were reclaimed by a massive sea wall 2,000ft long and 18ft high containing 15,000 tons of concrete. Work began in February 1902 under contractor J.& M. Patrick of Wandsworth; although the incomplete wall was breached in two places in January 1905, the area between the wall and the cliff was filled with 135,000 tons of shore sand and the Kingsway Gardens and promenade and shelters laid out by 1906. Lady Henderson cut the ribbon to open this new addition to the resort on 12 July 1906; (earlier in the day she had cut the first sod for Immingham Docks). The Boy with the Leaking Boot statue which stood in the Kingsway Gardens (but now in the Town Hall away from vandals) was presented in 1915 by John Carlborn; the original is in Stockholm, but there are also replicas in Fresno, California and Winnipeg.

South beyond the Kingsway, the local Council purchased 36 acres of seafront from Sidney

Sussex College in 1902 (this was the northern limit of the 13th century dune line), and a seawall erected in 1912; a further 230 acres were bought from the Marquis of Lincolnshire in 1921. Here the Bathing Pool was laid out in 1925, the Boating Lake in 1928 and the sea defences completed by building the Marine Embankment from near the sewage works across the Buck Beck saltmarshes to the dunes at Humberston Fitties in 1930. All amenity development has been behind the sea defences—the Olympia sports stadium (re-opened as the Winter Gardens in 1946), caravans and zoo at the southern end, while on the northern promenade are the candyfloss and slot-machine amusements euphemistically known as Wonderland. There before World War I was the spidery Warwick's Tower with its spiral lift and revolving observation platform, Hancock's Palace of Pleasure, the Arab village, a tortuous fairy river, a bicycle railway, swings and roundabouts, coconut shies and Aunt Sally, a water chute into the sea, helter-skelter, the 2d Switchback built on the beach, the Joy wheel (if you don't laugh you get a prize) and the Figure Eight railway.

Skegness on the other hand was gaining land from the sea almost from the inception of the resort. Skegness itself was a new town, built on land of the Earl of Scarbrough through the initiative and energy of his agent, H. V. Tippet. He had seen the possibilities from the beginning, and the first few plots were laid out in 1873; it was he who persuaded the GNR to increase both excursuions and ordinary trains, especially from London, and to run trains on Sundays despite petitions from Skegness and Wainfleet that they led to 'Sabbath desecration, dissipation and vice'. Although the grand plan (never completed as such) was not drawn up until 1878, 96 acres of farmland north of the old village street had been taken from the occupiers (chiefly William Everington); sites were being offered in 1876—reckoned to be the year of birth. That year plans were exhibited at railway stations in Nottingham, Derby and Leicester (where there was much interest) Boston, Lincoln, Grantham, Peterborough and Grimsby, Sheffield, Leeds, Bradford and Doncaster (Tippet was keen to interest Yorkshire people) and King's Cross. (Plans for a second railway, direct from Lincoln with a terminus south of the Vine Hotel, could have re-orientated the catchment area of Skegness.) The Earl visited Skegness in April 1877 and again with the Countess in July and August 1878, putting up at the Sea View Hotel, when the sea wall was being built (of Magnesian Limestone from his own quarry at Roche Abbey in South Yorkshire at a cost of £2,700 and lined with cast iron railings by Hill & Smith of Brierley Hill). The purpose of the sea wall was to carry the Parades, it being the landowner's responsibility to provide roads and services to encourage builders to buy plots or lease them at a low rent.

The idea of wide tree-lined avenues was Tippet's, and the estate controlled the building style, and line and type of foreshore development. Notices on the sands prohibited 'booths, stalls, swings, roundabouts, coconut sheets, shooting galleries, Aunt Sallies, or other erections of any description whatsoever' and that 'no hawkers with trucks, barrows, stalls or baskets, for the sale of fish, fruit, sweets or any other article whatsoever will be allowed on North and South Parade.' The plan included over six acres for laying out the Pleasure-Grounds (now Tower Gardens) on Grand Parade where the coal yard had been; the railway company paid Lord Scarbrough a farthing for every excursion ticket issued to Skegness so that a ticketholder could enter free. The opening grand fete was in May 1878. The gardens, with paths, trees and ponds, were laid out 'with considerable taste'; behind a 'commodious' pavilion were 'a skittle alley, water closets and lavatories, the last named being fitted up in good style, especially on the side set apart for the ladies'. The Sea Water Baths on Scarbrough Avenue opened in 1882 (at a cost of £4,000) with 'Tepid Sea Water Swimming Bath 6d, Private Hot & Cold Sea Water Baths 1s, Turkish Baths and Shampoo 2s'. No need at all to brave the breakers and breeze on the beach.

Between Scarbrough Avenue and Sea View Road was The Jungle, a children's paradise of scrubbed-up sandhills and marshy hollows; at the turn of the century part was made into The

Park by the Earl. (The town north of Scarbrough Avenue was not developed until the 1920s, by T. L. Kirk, a Nottingham solicitor, and J. H. Canning.) In the dune edge along North Parade was George Burrows' Chalet Theatre (where is now the Imperial Hotel) and the Switchback, built in 1885 but superseded by the Figure Eight railway (built by Thompsons' Patent Gravity Switchback Railway Co.) on the new foreshore beyond the Parade in 1908 (dismantled 1970). The dunes of The Jungle had other attractions: 'paterfamilias goes there to read his morning paper in peace, and thither the devoted swain conducts his Phyllis', the stile entrances from North Parade 'being handy lessons in gallantry for gentlemen'. Low dunes had been accumulating in front of the sea wall from the time it was built. This allowed the Marine Gardens to be laid out south of the pier by Lord Scarbrough in 1887-8. The foreshore beyond was of parallel sandy ridges and wet (and sometimes muddy) runnels or swatches. In summer the children dug sandy bridges or laid planks across the swatches, charging trippers using them.

Although there had been a marked decline in the number of excursionists in the mid-1880s, due to depression in the industrial catchment areas, by 1890 Skegness was described as 'the noisiest and most crowded of the Lincolnshire sea-side places, except Cleethorpes, invaded every day during the summer by an enormous number of excursionists from the midland counties . . but . . . after 6pm the place is again nearly emptied'. While the same Guide said that Skegness was 'not much to be recommended for quieter visitors', the curative qualities of the bracing breezes, 'in high repute among the medical men of London and the Midlands for its good effects in cases of chest complaints and rheumatism', induced the building of the Nottingham & Notts Convalescent Home for Men (1891) and for Children (1892-3) at Seathorne (with funds provided by Sir C. Seely), Lady Scarbrough's Home for Women & Children at Seaview (patients paid only 5s a week), the National Deposit Friendly Society's Convalescent Home on North Parade (1927; it became a Town Hall in 1964) and the Derbyshire Miners' & Friendly Societies' Convalescent Home at Winthorpe (1928). The Glass Shelters on North Parade were declared to be a 'boon to visitors of delicate health'. The dunes north and south of the town were excellent for golf courses—established at Seacroft 1892-5 and the North Shore course in 1911—and hotels sprang up along the front including the Seacroft Hydro (Hotel) built in the red brick and terra cotta Domestic style of the 18th century in 1908-9.

The most famous seafront feature of Skegness is the 56-feet Clock Tower, erected 1898-9 by public subscription (£550) to commemorate Queen Victoria's Diamond Jubilee, and 'unveiled' by the Countess of Scarbrough on 11 August 1899. The architect was Edmund Winter of Liverpool and it was built by W. H. Parker & Son of Boston. Acknowledged to be of 'graceful design', it was hoped that 'time's mellowing hand will soon tone down the somewhat too glaring effect of the whole'.

Experiences of day trippers varied in the 1890s. The 623 customers and friends of Jackson & Son, clothiers and outfitters of Horncastle, who went on a long half-day for 1s 6d return on Wednesday 26 August found themselves as the only trip. The same week of 1896 was a record, with a Methodist chapel being used for sleeping and hundreds had to find lodgings in Wainfleet and Burgh. The problem then was the unannounced arrival to stay for a few days of 4,000 members of the Order of Sons of Phoenix from London. Numbers of excursionists to Edwardian Skegness built steadily from 226,887 in 1902 to 321,260 in 1907. On 13 February 1908 the GNR purchased for £12 from artist John Hassall an oil painting of a smiling, pipe-smoking fisherman bounding along a sandy beach. The words 'Skegness is so Bracing' were added for a poster to advertise trips from King's Cross for 3s return (for a 3-hour journey). Numbers increased to 356,409 in 1910, and in 1913 the total number of passengers arriving at Skegness exceeded three-quarters of a million. In the eight weeks up to 4 August 1914 the figure was 407,000. It was never the same again, even though the waves on the shore became those of the North Sea rather than the German Ocean.

ABOVE: Waltham Sunday School outing to Cleethorpes, 4 August 1914.
BELOW: Trippers fill the promenade and sands at Cleethorpes.

LEFT: Cleethorpes cockler, 1911. RIGHT: Levi Stephenson—the man with the big nose. CENTRE: Empire Theatre, Cleethorpes 1906. BELOW: A family paddle at Cleethorpes—showing a daring amount of leg, 1908.

ABOVE: View of Cleethorpes front from the revolving tower, 1905, CENTRE: The Promenade and Switchback, Cleethorpes c1905. BELOW: High Cliff, Cleethorpes 1861.

ABOVE: Cleethorpes cliff and pier 1880. CENTRE: New promenade 1906.
BELOW: New promenade and beach amusements, Cleethorpes 1911.

Sea View Road, Cleethorpes; ABOVE: in 1904, and BELOW: 1905.

LEFT: An Edwardian promenade at Cleethorpes. RIGHT: The Boy with the Leaking Boot. BELOW: Cleethorpes Boating Lake in the old dunes, with new marsh and dunes on the foreshore. (GET)

Bathing Pool and Kingsway 1955. (GET)

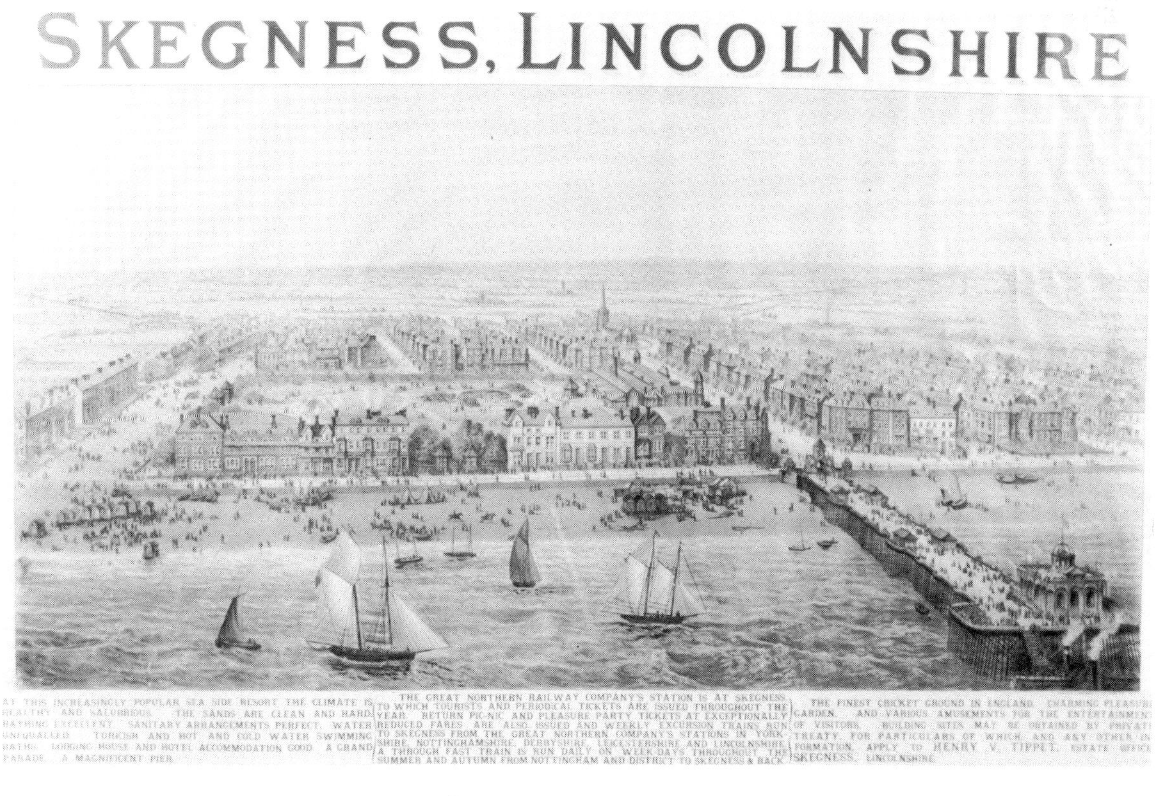

LEFT: Gipsy Roma. RIGHT: Kiddies Corner, Cleethorpes was for grown-ups too. BELOW: Artist's view of what Skegness might have looked like had the full plan—shown overleaf—been completed. (LLS)

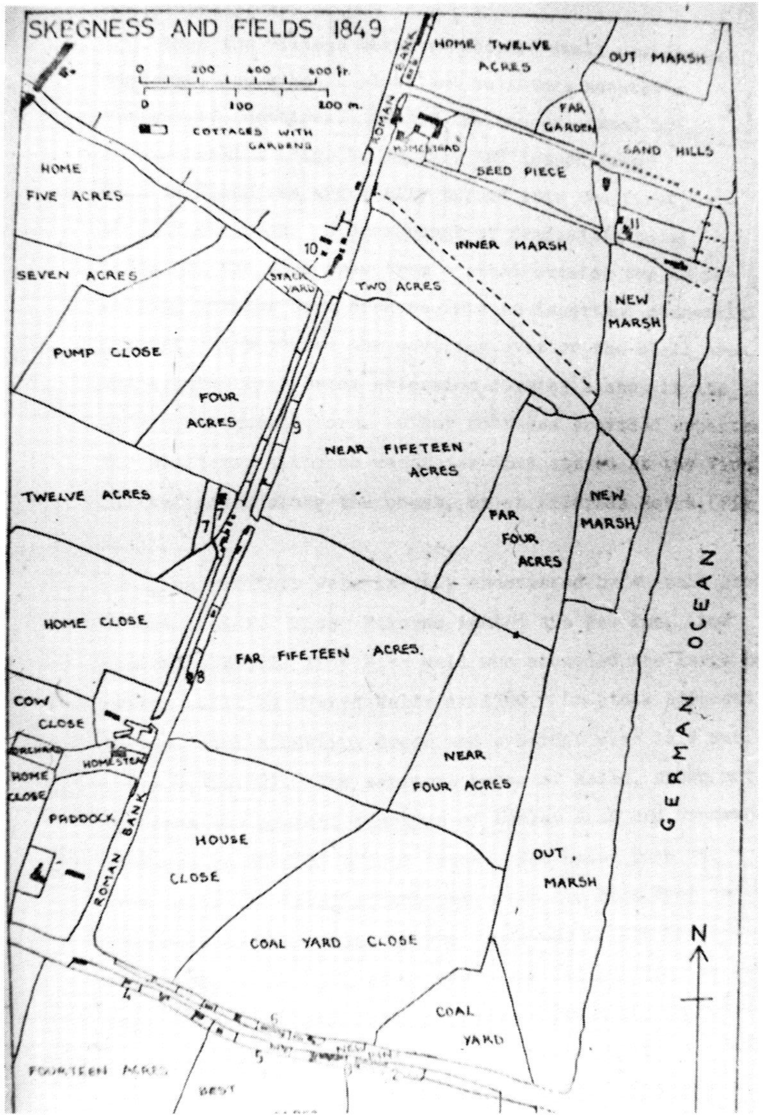

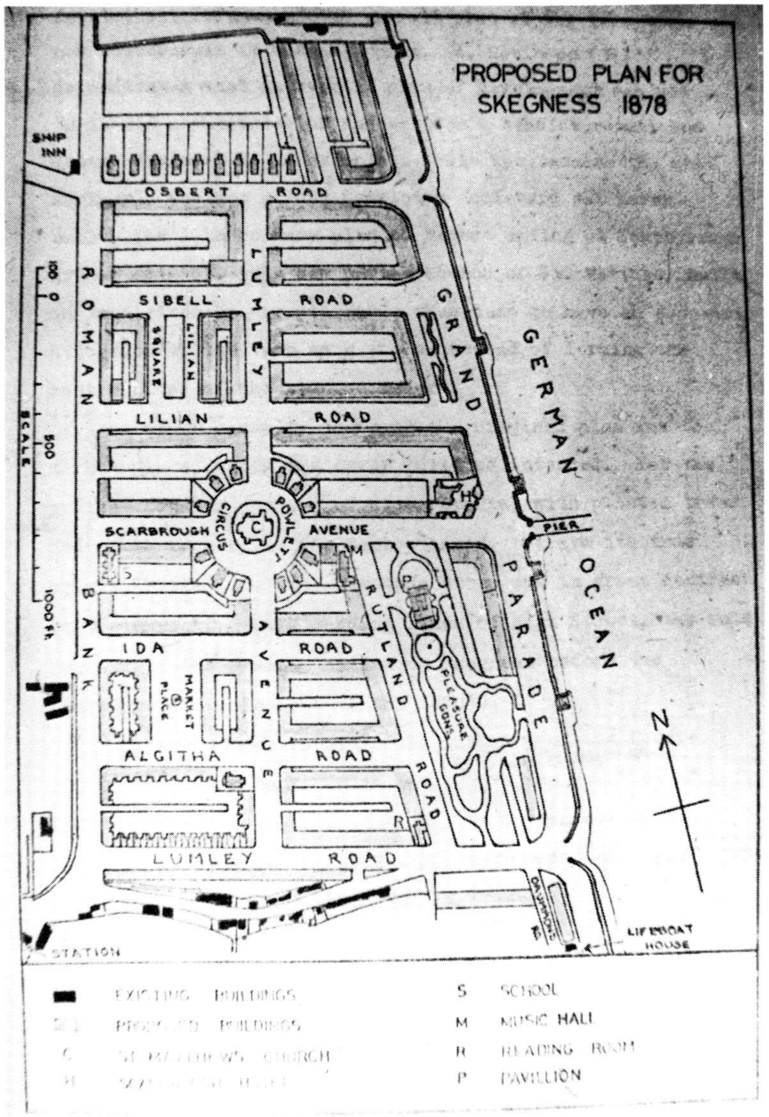

LEFT: Skegness in 1849. RIGHT: The proposed plan for Skegness.

ABOVE: George Dunkley's Sea View Hotel. CENTRE: Grand Parade before the Clock Tower was built. BELOW: Sea Water Baths, Scarborough Avenue 1902.

'Unveiling' the Clock Tower, 11 August 1899. INSET LEFT: Tower Gardens, Skegness 1905, and RIGHT: Figure 8 and model railway, 1926.

ABOVE: South Parade, Skegness 1900. CENTRE: The Ship Eliza—Joe Wingate's Whale Museum (2d to see the skeleton); destroyed by storm in 1911. BELOW: Nottingham Convalescent Home, Seathorne.

ABOVE: National Deposit Friendly Society Home, later Skegness Town Hall. CENTRE: Derbyshire Miners Convalescent Home, Winthorpe. BELOW: W. B. Thomas painting of Skegness showing the new Seacroft Hydro.

ABOVE: A bridge over a swatch keeps your feet dry after a high tide.
BELOW: B. C. Hucks flies George Burley and daughter of Skegness in his 70hp Bleriot, 1913. (WK)

ABOVE: The Skegness Sun Castle or Solarium. CENTRE: Sun Castle trio: Florence Andrews piano, Carole Chester violin, Elsie Cordell drums. BELOW: The new Orchestral Piazza (now the Embassy) on Grand Parade, Skegness.

Beside the Seaside

In December 1921 Skegness Council Surveyor R. H. Jenkins presented his master scheme for the development of the accreted foreshore which had been purchased from the Earl of Scarbrough for a modest £15,590. The plan showed a boating lake, bathing pool, putting and bowling greens, tennis courts, refreshment rooms, formal gardens and shelters. The Council approved the plan, not without misgivings about possible costs and uncertainties about the pattern of post-war holidaytaking, but it was Jenkins' enthusiasm and organising ability that created the splendid seafront amenities of today. The same year, 1921, a young man called Billy Butlin had arrived back in England from Canada with £5 in his pocket. Within a few years, travelling showman Billy Butlin had set up his hoopla and three other stalls in The Jungle where the County Hotel now stands. He came to Skegness to relieve the trippers, who now also began to arrive in charabancs (or 'chonkabonks'), of their money. He also had a good eye for the best site—opposite the Clements Entertainers in Happy Valley and between the town and the amusements (roundabouts, swing boats, skittle alleys, aerial ride and Figure 8) on North Parade.

The Boating Lake was opened in 1924 and was such a success that it was extended two years later. The old Marine Gardens were transformed into bowling greens, tennis courts, an artificial rock garden 'which might have been fashioned by Puck and the fairies', and the open-air 'New Wonder Pool of the East Coast' was opened on Whit Monday 1928. The following month Billy Butlin suggested a large amusement park south of the pier. It was based on the law of the seaside landlady: out after breakfast and don't be late for lunch, back to the beach in the afternoon and in for tea; which left the question, where to spend the cash, particularly if it was dull or wet. He obtained a long lease from the Council and was on the high road to fortune when it opened in 1929. He introduced the Canadian dodgems to Britain at Skegness (including water dodgems) and the seaside zoo. The Orchestra Piazza (now Embassy Ballroom) opened in 1929 and the Sun Castle or Solarium in 1932. That year at the southern end of the new developments, and facing the open beach, appeared a row of wooden chalets, on the same style as the former bathing machines, but static; the wheel had come nearly full circle.

Wheels of a different sort had been turning for some years on the firm sands of Skegness beach. Motor cars with names like Alldays & Onions, Argyll, Buckboard and Darracq competed with Humbers, Raleighs, Minervas, Oldsmobiles and Daimlers on a straight mile course (with wire netting over soft sand patches) at a first meeting organised by the Notts Automobile Club in 1905. After World War I, when a prototype Lincoln tank had been tested on the beach, annual Motor Races started again in 1923, attracting names like Malcolm Campbell and Raymond Mays. In 1929 the course was changed to a 2km oval with exciting hairpin bends at each end and transferred to the north end of the beach. Competitors came from as far afield as Brooklands and Leeds, driving Bugatti, Francis, Salmson, Frazer-Nash, Talbot, Beljap, Bentley, Vauxhall and Mercedes-Benz cars. In 1930 the Boston & District Motor-cycle Club held its first race meeting on the sands at Chapel St Leonards.

Cars and motorcycles now competed with the train as a means of getting to the seaside. The record Whit weekend 1930 was a clear demonstration of R. H. Jenkins' foresight, helped by 13 hours of sunshine on each of the Sunday and Monday. On the Monday 70-80,000 people came in 31 excursion trains, and up to 10,000 charabancs, cars and motorcycles—some arriving before many staying visitors had finished breakfast—at peak rates of 1,200 an hour on the Burgh Road and 850 an hour on the Wainfleet Road. Over 10,000 people went to the Bathing Pool (only a quarter to bathe) and all 130 boats on the Boating Lake were let by 10am.

A new era in seaside holidays began when the Butlin Luxury Holiday Camp at Ingoldmells opened in Spring 1936. Representing an investment of some £100,000 and a test of judgement, its motto was, and still is, 'Our true intent is all for your delight' (from *A Midsummer Night's Dream*). The former farmland on which it was built lies between the 'Roman' Bank and the sandhills, the strengthening of which created a large hole which Butlin turned into a boating lake. A water supply was obtained by boring some 80 feet through the marine silts and boulder clay into the underlying chalk. The site had appealed to him, because of its road access and proximity to the beach, one day when travelling from Skegness to his amusement park at Mablethorpe. The fact that Butlin had such a park there exemplifies a major difference between the development of Mablethorpe and of Sutton, although both were hemmed in by defensive sandhills and later sea walls.

That both Mablethorpe and Sutton offered a healthy, invigorating climate was never in doubt, and both claimed, in varying degress of accuracy, firm and safe sands. The lifeboat at Sutton in 1866 was purchased by subscription among the working classes of Birmingham. In the 1870s annual sand races and a regatta were held at Sutton, which could be watched from a promenade 'with covered ways and comfortable seats' opened in 1876 (replaced with a quarter-mile sea wall with asphalt promenade and carriage drive in 1885—a relic of spa elegance). It was suggested that Mablethorpe might care to follow suit. What Mablethorpe did have, about half a mile south of the village, perched on a seawall in the dunes, was the Lincolnshire Sea Side Convalescent Home opened in July 1871 (at a cost of £3,800). The idea of Miss Emily Anderson when she started collecting subscriptions in 1866 was 'to give the working classes some important sanitary advantages which would otherwise be beyond their reach'. She was not very successful but Rev Frederic Pretyman of Great Carlton took up the work. James Fowler of Louth designed the Home for 53 patients (who paid their own travel expenses and 5s a week) with separate accommodation for males and females on a plan approved by Florence Nightingale. A bath house was added in 1875. (In the dunes north of the village the Leicester Poor Boys' & Girls' Institute opened their Summer Home in 1906.) The Eagle Hotel was established around the same time as the Convalescent Home, and provided, as did the Book in Hand and the Jolly Bacchus, transport to meet trains at Alford. In 1876, when John Simons had made extensive alterations to the Jolly Bacchus, there were only four lodging houses in Sutton compared with 18 in Mablethorpe (and there were still only 10 in Skegness where the railway had been open for three years).

The opening of the Louth & East Coast railway to Mablethorpe on 16 October 1877 (too late for that season's trippers) was spectacular: the first train came in backwards and smashed through two crossing gates. The village was already developing in a relatively unco-ordinated fashion (compared with Skegness) in anticipation of 'numerous passengers . . . from large manufacturing towns of Nottingham, Derby, Sheffield and Manchester . . .who will be in close communication with this charming and salubrious watering place'. The Book in Hand was entirely rebuilt by John Thorsby in 1880 and by 1882 there were 29 lodging houses in the village.

There was a flurry of plans and activity in both Mablethorpe and Sutton in the 1880s. Plans for a branch of the Louth & East Coast railway to North Somercotes and Cleethorpes of 1872 and 1877 had come to nothing, as had a second line from Alford to Mablethorpe in 1873 and a tramway from Alford to Chapel St Leonards and Skegness in 1883, but attention began to focus on Sutton. A steam tramway along the road from Alford to Sutton opened in April 1884 and the same year the North Sea Fisheries Harbour & Dock Co. Act was passed. This would have created a large harbour at Sutton a quarter of a mile south of the pullover: two massive concrete piers with a lighthouse to enclose a 14½-acre refuge and a lock entrance to a 10-acre inner harbour. (The idea was revived in 1891 with plans for a railway from Warrington to Sutton, which got no further than Lincoln, and it died when work started on Immingham Docks in 1906.) In anticipation of the docks, the Sea View Freehold Land Society was formed in 1883, mainly by Nottingham businessmen, to develop 75 acres north of Sutton High Street. The result was The Park which enhanced the better standing of the place, as did the choice of Sutton on Sea rather than Sutton le Marsh as its name (but not officially confirmed until 1956).

In response the Mablethorpe Improvement Committee built a pavilion on the sandhills north of the pullover in 1883 (where a short wooden Esplanade was erected by 1907) and Mrs Catherine Benson gave another pavilion on the south side in 1884. The Sutton & Willoughby railway opened in October 1886, another piece of poor timing (and effectively killed off the steam tram service, which struggled along to 1889) but the Sutton and Mablethorpe termini were not joined until 14 June 1888. That was a success, with 3,000 bookings from Mablethorpe to Sutton in two months. In the second half of 1889 some 40,000 people used the line.

In 1890 Sutton was 'not so crowded and noisy as Skegness, or even Mablethorpe' with broad, firm sands 'well suited for children'. Board and lodging could be had with Rebecca Simons at the Bacchus Hotel for 5s 6d a day, servants 4s 6d, or private rooms at 10s-12s a week, with bathing machines at 6d per person (except Sunday) and there were three tennis courts and a 'spacious Billiard Room, with a Burroughs & Watts Table, fitted with new patent Eureka low fast cushions'. Mablethorpe in 1890 exhibited 'hardly any of the energy for creating popular attractions which had been conspicuous at Skegness' and bathing from the shore was 'not very safe, owing to the clay-banks, in which there are some deep holes'. (A 22-year old Manchester man had been drowned there in 1877 when swimming in a receding tide.) 'Mablethorpe frankly admits it wants the tripper; Sutton on Sea does not.' Sutton had gained the reputation of being a 'select holiday resort' with absence of 'the usual conglomeration of amusements', although around 1898 the Sutton on Sea Amusements Association was formed to advertise and promote the place (but with no Sunday work—the Methodist influence of Benjamin Simons).

Perhaps it was the influence of the GNR's 3s return from King's Cross before World War I, which brought 92,000 people to Mablethorpe in 1906 (including 5,400 trippers on August Bank Holiday) that persuaded the Urban District Council to acquire all their sandhills in 1914. When D. H. Lawrence returned to 'the flat dree coast' in 1926, he found Mablethorpe 'rather nice, a quite common seaside place, not very big, with great sweeping sands that take the light, and little people that somehow get lost in the light, and green sandhills'. Interwar excursion traffic to Mablethorpe reached a peak of 166,000 in 1936; by this time Butlin's amusement park was on the levelled dunes where Miss Benson's pavilion had been (there is nowhere else for permanent structures unless inland of the sea defences). An attempt to build a swimming pool over the tidal basin of the drain outfall (where 'Professor' Hobson Bocock gave daily swimming lessons in Edwardian times) was unsuccessful in 1930. Excursion traffic never recovered after World War II; there were 136,000 passengers in 1959 but the following year the Louth-Mablethorpe line closed, and numbers dwindled to 6,500 to Mablethorpe and 12,500 to Sutton in 1963. Seven years later the towns lost their rail link.

What might have happened had the 1908 dream of a Garden Village by the Sea called Woldsea become a reaility? The Huttoft Town Planning Syndicate Ltd, whose local estate agent was none other than Benjamin Simons, published from their London office an excessively effusive and profusely illustrated prospectus of the 1,135-acre project with a two-mile sea frontage from Anderby Creek to Sutton golf course, centred on Huttoft Bank. The 'picturesque' sandhills were to be converted with flowering shrubs and trees into 'charming Lees'. The Great Northern Railway passed within a mile and all roads led to Woldsea, it was claimed, and would enter through a facsimile mediaeval gateway. A villa would cost £2,000 and a bungalow £150. 'Freak building' was to have no chance and there would be no 'jerry built lodging houses'.

Scorn was directed at Skegness—'a London suburb planked down in a hurry on the ocean's marge', and at Sutton where during the season were 'nigger coons, battles of flowers, tennis tournaments and other dissipations of the kind'. There was to be no pier or fashionable promenade—just quietness and seclusion in 'a watering place on entirely new and superior lines which will appeal to the better middle class', with an emphasis on all 'health-giving sports'. In the years up to 1914 the titled, the well-to-do and about-to-retire colonials came to Lincolnshire to inspect the site, look at the plans of thatched cottages and stockbroker Tudor villas, and went away again.

The first priority of the planners was a 'tastefully designed and fitted up' hotel—The Links. A hotel was already in existence in Sandilands by 1909—The Grange. This became the Grange & Links in the early 1920s—perhaps the only tangible link with the dream that faded. Except that some of those who came to see the empty fields and sands of Huttoft did return to retire—if the semi-colonial style of some of the dwellings in Sandilands and Skegness, both adjacent to golf courses, are anything to go by.

Skegness 'New Wonder Pool of the East Coast'.

ABOVE: Racing car meeting on Skegness sands 1905. LEFT: Seafront 'toastrack' 'bus service at Skegness 1920. RIGHT: Billy Butlin publicises his new dodgems. BELOW: Waiting for a sail—Skegness 1910.

ABOVE: Skegness boarding house party 1929. (DL) BELOW: Butlins Holiday Camp, Ingoldmells from the air 1965. (GET)

ABOVE: Skegness from the air in 1965 showing the extensive foreshore and amenity developments beyond the original Parades. (GET) LEFT: Holmdene, Drummond Road, Skegness where my granparents lived and where I spent my first holidays. RIGHT: Derbyshire House, South Parade, Skegness—a typical seafront boarding house.

ABOVE: A seaside holiday at Sutton in the 1890s—a Nainby study. (WERH) BELOW: Daughters and donkeys at Sutton in the 1890s—a Nainby study. (WERH)

ABOVE: Dressed for the seaside in the 1890s—a Nainby study at Sutton. (WERH) BELOW: Bathing styles at Sutton in the 1890s—a Nainby study. (WERH)

ABOVE: Children's Special Service Mission for Sutton and visiting children held annually at Sutton, led here in 1896 by J. F. Allen of Nottingham. (WERH) BELOW: Beach Hotel, Sutton in the 1890s—a Nainby study. (WERH)

ABOVE: Lincolnshire Convalescent Home, Mablethorpe: original architect's drawing by James Fowler 1871. BELOW: Eroded sandhills behind the sea wall at Sutton reveal the clay core of the old sea bank, c1890.

ABOVE: Sunny days in Sutton sandhills in the 1930s. CENTRE: Wooden sea defences and chalets at Sutton, 1930s. BELOW: Having a go on the Aerial Ride at Mablethorpe—which was little more than a pulley with handles running down an inclined wire, but with a safety net.

ABOVE: In Edwardian times the Hall family from Louth and relations lived for a month in bell tents on the beach at Mablethorpe—and had trouble with high tides. LEFT: Things had changed by the 1960s. RIGHT AND BELOW: The Swimming Basin at Mablethorpe—the drain outfall—where 'professor' Hobson Bocock (black tights and moustache) gave swimming lessons daily.

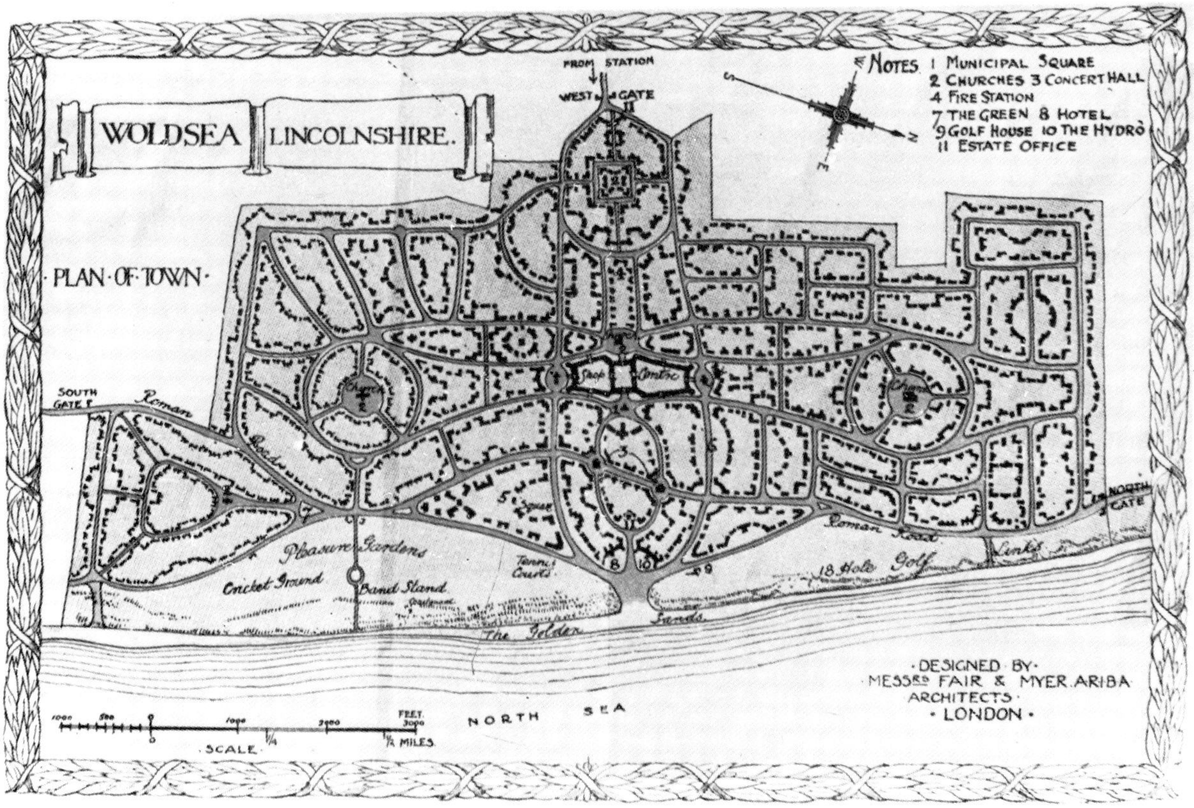

ABOVE: Mablethorpe railway station crowded with arrivals and departures in the days of 3s excursion fares. BELOW: Woldsea, the town plan. (RF)

Woldsea: ABOVE: a planned hotel, CENTRE: a pair of houses for £1200, and BELOW: a suggestion for The Green. (RF)

ABOVE: Cleethorpes pier and approach from the cliff top, 1890s. BELOW: Cleethorpes pier pavilion fire 29 July 1903.

Piers and Pierrots

. . . that sea-monster millipede, the pier,
Puts out from shore to please a giddy throng.
H. D. Rawnsley (1887)

The weather-beaten poster announced Mike and Bernie Winters topping the bill at the Pier Theatre. Little did they know when the show closed on 3 September 1977, that it could prove to have been the last on Skegness pier. Just beyond the poster was a mass of twisted girders and two yawning gaps with only the central section and its two shelters and the pier head remaining. It was the morning of 12 January 1978, and it looked like the end for the 97-year old pier.

The previous evening a northerly gale had coincided with a high spring tide. Massive waves, driven in from the north, had lifted the wooden decking, twisting the supporting girders until the pounding seas finally sheared off huge sections. The section between the pier head and the shelters broke up first. So powerful was the sea that parts of the pier structure, including sections of iron seating, were found as far south as the spit at Gibraltar Point, four miles along the coast. Although much of the timber decking was recovered, it will not be possible, except at exorbitant cost, to reproduce the Victorian architecture of those sections swept away. Equipment from the 1000-seater theatre stranded on the pier head was salvaged between tides in the hope that it might be possible some day to join up again the pier's shattered remnants. Solutions including cable cars or a monorail have been suggested, but it seems that one of the main attractions of Skegness—a walk to the end of the pier—is a thing of the past.

The first tentative step towards pier building along the Lincolnshire seaside was taken in 1867 when the Cleethorpes Promenade Pier Order was confirmed, but it was five years before a company was formed to promote construction. The cost, £10,000, was raised in £10 shares. Chairman was Edward Bannister, coal merchant, and most shareholders were local. Construction began straightaway of the 1,200 feet iron-piled pier, with an elevated entrance from the cliff-top gardens, and a concert hall at the seaward end. Contractors were Head Wrightson of Stockton on Tees, the major pier builders of the period. At the grand opening on August Bank Holiday 1873, 2,859 people paid the special admission charge of 6d to promenade over the briny. Thereafter the toll was 1d, and 37,000 persons took advantage in the first five weeks. Clearly the pier was as popular as the other 34 already built in Britain. So popular indeed that a landing stage was added at the pier head in 1875 to facilitate sea trips to Spurn Point on the Yorkshire coast.

The Manchester, Sheffield & Lincolnshire Railway, whose excursions (from Rotherham and Sheffield at 6s return first class or 3s in 'covered carriage', Retford and Gainsborough at 5s and 2s 6d) brought in most of the patrons of the pier, and who were in the process of encasing the crumbling cliff, leased the pier in 1884 at an annual rent of £450. Two years later the landing stage was damaged by fire and then the concert hall was similarly destroyed on 6 July 1903. This did not deter the railway company from buying the pier for £11,250 the following year; it replaced the hall with cafe and shops, and built a new pavilion complete with stage and dance floor part way along the pier in 1905, which was opened on 10 June by a Grand Promenade Concert with contralto Ada Crossley. 'The floor, from the character of its construction, is admirably adapted for dancing, and in the morning, in the afternoon and in the evening the cadence of a waltz, or the ring of the lancers, floats way over the sands and sea. Under the glass-covered verandah, hung with flowering plants, young men and maidens meet.' The installation of heating later allowed the partaking of these enjoyments in the winter. With the extension to seat 600, the pavilion provided regular concerts of a high standard; after 1923 there were celebrity concerts on Sundays only, although it had become the home of the highly acclaimed Cleethorpes Music Festival from 1909.

In 1935 the pier was purchased from the LNER (as it had become) by the local authority under the Cleethorpes UDC Act of 1928 for £10,000 (the same as the construction cost 68 years before). Three years later the entrance was altered to give access from the promenade. As an anti-invasion precaution the pier was deliberately breached in 1940, only to find after the war that the estimated £26,000 cost of putting the whole into good repair was prohibitive, and the seaward section was demolished leaving a rump of 335 feet. Two hundred tons of timbers went to rebuild Leicester City's football grandstand and most of the remaining materials went into the construction of Wonderland on the North promenade.

Just over four years after the opening of the Cleethorpes pier, on 22 November 1877, a group of Skegness businessmen met in the New (or Hildred's) Hotel to discuss how their town could acquire the necessary status symbol of a pier. The Skegness Pier Company was formed and a £50 premium offered for the best design. The winner, chosen from 44 entries, was by Clarke & Pickwell of Hull and was the first in Britain not to follow the Oriental tradition. By May 1879 a provisional order of approval had been received from the Board of Trade and the engineers, again Head Wrightson, commenced work at the end of the year. The pier was erected on hollow cast iron piles screwed 30 feet down into the sand, mud and clay. Into these were bolted iron columns with decorated tops of scallop and leaf foliage design. Resting on the columns were two sets of girders separated by lattice girders of wrought iron. On top rested thick beams of Oregon Pitch Pine across the pier with secondary joists of Baltic Redwood. Finally the decking was of Jarrah timber; and it is significant that it is called a deck and not a floor. Wooden seating ran the length of the pier on each side, the cast iron back-rest interspersed with shields showing the initial letters of the Skegness Pier Company and the date of foundation of the pier (1881), and decorated on either side with dolphins. The same dolphins were used in the arm rests. The gas supply for the lighting at the pier head was carried in the two inch handrail above the seating on the south side.

The cost of the 1843-foot pier, the fourth longest in Britain, was £20,000, most of it coming from debentures bought by the Trustees of Lord Scarbrough's estate. The pier was approached from the promenade by a lumpy Gothic archway which would not have been out of place at a cemetery, but remained until 1936. Either side of the entrance was a kiosk, one occupied by a barber and the other by palmist Madame Elsa; they were removed in 1920. The opening ceremony on 4 June 1881 was performed by the then Duke of Edinburgh, and Edinburgh Avenue off the Grand Prade was named after him. Within a couple of years the pier was damaged by an overturned vessel being swept from the north through the girder work, and in May 1894 it became necessary to protect the pier head with granite as the sand and clay around the piles had been scoured away by wave action.

Steamboat trips from the pierhead landing stage, which allowed embarkation three hours either side of high water, had started in 1882, and the following year the Skegness Steamboat Co was formed by local businessmen to provide trips round the Wash. By 1883 the paddle steamer *May*, one of the largest on the east coast with a capacity of 225 persons, operated daily except Sundays. A one hour cruise cost 1s, or 1s 6d for a 2½-hour trip to the Lynn Well lightship including a tour of the vessel. The most popular trip was to Hunstanton, leaving Skegness at 8.30am and returning about 8pm at a cost of 3s plus pier fee at Hunstanton. Paddle steamers such as the *Scarborough, Yorkshire Lass, Buster, Spendthrift* and *Privateer* brought trippers from the Wash ports, returning them at night. In the late 1920s the North Sea Steamship Co operated a triangular Skegness-King's Lynn-Hunstanton service, and in 1929 as many as 7,000 people were landed on the pier from pleasure boats. As the nature of the foreshore changed, with increased accretion of sand there was insufficient depth of water for pier landings and passengers were transferred to small boats from the beach.

A small fire in August 1895, which burnt a portion of the decking and saloon drew attention to the 'inadequate extinguishing apparatus', and 28 fire buckets were placed along the pier and round the pier head. Later an asbestos pipe was installed under the seating on the south side and water would be pumped from the boating waterway which passes under the pier at the beach end. In 1905 it was decided to provide a gents' WC at the pier head. An expert, called in to advise, suggested that the south-east corner be chosen so that the prevailing wind would blow away any smell. The pier suffered a severe disaster on 21 March 1919. The motor schooner *Europa* was on the way from Amsterdam to Wisbech in ballast when she was caught in a north-easterly gale and anchored on the Outer Middle. The previous evening's tide had driven the ship on to the beach and left it dry, but the skipper chose to ignore warnings from Matthew Grunnill, coxswain of the lifeboat, about the seriousness of his position. Soon after daybreak a crowd gathered to watch the vessel on the incoming tide. The anchor, as predicted, failed to hold. The sweep of the tide threw the stern of the ship under the pier midway between the two sets of shelters. With every lift of the waves the decking fractured and sheared off. The continued battering demolished 150 feet of the pier and the receding tide left the vessel lodged in the gap. A temporary gangway was erected for just over £1,000, but it was 20 years before the break was finally repaired by local engineer Chris Gutteridge at a cost of £373.

When the pier entrance was altered in 1936-37, the original balustraded ramps were demolished and a new entrance in 20th-century Odeon style was erected with cafe and shops incorporated. During World War I the pier had remained open, to reduced business, but during World War II it was occupied by the RAF at the pier head and the Navy at the entrance, with the Army inbetween. In 1946 a steel and glazed frame was erected over the concrete framework which had replaced the original ironwork between the entrance and the beach steps (installed 1929-30). This became the penny arcade and provided much of the pier's income. In 1971 the entrance was altered yet again. The flight of gracious steps leading to the deck was removed and the building brought to ground level. A characterless mass of glass and concrete contains an amusement arcade, prize bingo and luxury variety bar. A far cry from 'What the butler saw'. When the pier company had been in financial difficulties in 1914, Lord Scarbrough bought the remaining shares; he sold the pier to a syndicate of four local businessmen in 1964. The majority of shares of the pier, or what is left of it, are now owned by R. G. Mitchell Ltd.

The business of a pier is at both ends. The building at Skegness pier head was variously the Concert Hall, the Saloon, the Pavilion and the Pier Theatre as the auditorium was extended to incorporate open deck and kiosks (such as those of phrenologist G. H. J. Dutton and the Camera Obscura of Mr Davies). The pier show of August 1890 presented ventriloquist Captain Slingsby, banjoist Walter Howard, Signor Isidore Corelli in 'Silent Sorcery' and contralto Myfanwy Morris. Seats for such secular entertainment were 1s and 6d (including pier admission) but Sunday 'Sacred Concerts' were free. In the Concert Hall at the end of Cleethorpes pier in 1892 there was 'smart entertainment given by Mr W. J. Hunter, Miss Bayley and that able vocalist and decidedly clever harpist Mr Charles Clifford', while in the Gardens were the Snow Brothers Minstrels and the Railway Servants Brass Band. Skegness pier had the Band of the 16th Lancers in 1911, and in the decade before the First World War

Surtees Corne's Orchestra gave daily concerts on the Deck at 11am and 4pm, and there was 'refined variety entertainment' in the Saloon at 3pm and 7.30pm. In the 1920s there was Contri Tait providing music and Will Gardner's Skegness Entertainers, the Pier Confetti Fete and the Masked Carnival. In the post-war era there was the popular Jan Ramsden trio and 19 seasons of Ted Dwyer's 'Follies on Parade' running from Whitsun to October, while Oscar Rabin and his band were on Cleethorpes pier. Other attractions provided on Skegness pier have been roller skating (in 1907, 25 pairs of skates were bought for 'rinking' in the Saloon), and diving off the end of it. There was an 80-feet high diving board on the west side of the pier head from which Professor Connell performed his Great High Dive in 1898 and continued annually through the next few seasons. Later there was Babs DuBarry, and in the 1930s the one-armed Daredevil Leslie. After his exciting act, which included dropping like a stone in a sack, he walked the pier, dripping wet, shaking a box for impressed and gawping onlookers to pop in a penny, while he called out, 'Don't let your hearts be as cold as the water'.

Mablethorpe might have had a pier with similar attractions had there been local enthusiasm and money. When Hull solicitor Robert Galland was promoting the abortive Lincolnshire Marshes Railway (from Mablethorpe through North Somercotes to Cleethorpes) in 1886, he also issued a prospectus for a pier. A civil engineer was engaged and a site chosen, a short distance south of the pullover. Wealthy landowner Mrs Catherine Benson of Utterby near Louth, who also had a seaside home in Mablethorpe (now the Regent Cafe), agreed to sponsor the scheme but died suddenly, before it could be brought to fruition.

Pierrots had nothing to do with piers, except that sometimes they performed there. Of French derivation, their rise was rapid from 1891, soon replacing nigger minstrels who had been around for over 30 years. By 1905 Cleethorpes had Alf Wilson's White Star Pierrots, and the Imperials which included two girls—H. Dunville, Nita Carmen (Mrs Dunville), Edith Mayfield, Frank Clerk and Peter Parr—together with Miss Nellie Dyson's Jolly Mascots. A little later, in 1911, there were the all-male Ellison's Entertainers. The Alexandra Hall Theatre in Cleethorpes had opened in 1896, soon renamed the Empire, and now demoted to bingo. The Coliseum in the High Street opened as a cinema, was used as a boxing hall by the Manchester Regiment during the 1914-18 war, and then variously by the British Legion, as an auction room and a dance hall before becoming a multiple store in 1952.

The Skegness story, and to a lesser extent Mablethorpe's is that of the Clements family. It all began when father and three sons, Fred, Bob and Jack were working for the Great Northern Railway at King's Cross. Although Londoners failed to respond to 3s Sunday excursion fares to Skegness, Fred noted the annual influx of people from Nottingham, Leicester and Derby and decided to try his amateur pierrots, complete with ruffles, pompoms and skullcaps. He obtained a pitch on the sands by the Clock Tower, got a local builder to make a stage, took a week's holiday, and made a start on Whit Saturday 1902. Somehow they could not compete with Billy Marsh's Nigger Minstrels. The north-east wind blew and they managed only four shows. They were a flop. Then Fred recruited a company of young and good-looking men 'to attract all the girls from the Nottingham lace factories'—ventriloquist Johnson Clark, soloist Wally Young, comedian Chas Coverdale and his greatest asset, six-feet pin-up Frank Rainbow of Boston who sang, would you believe it, 'I love daddy, my dear daddy; he's my playmate, rain or shine; there's not a daddy in the world like mine'. The crowds came, and the Clements had arrived. The next season they had to move to the undeveloped dunes in front of the Jungle just beyond the pier, but so close that Fred could collect from onlookers on the pier using long bamboo poles with a net on the end. For that 1904 season they had to move away from the pier; Fred laid a path of railway sleepers from the pier into a hollow in the sandhills where the Sun Castle is now. With the exception of the war years, 1914-18, there they stayed until 1930 in what became known as Happy Valley.

At the same time Fred built a stage with canvas roof over the audience for evening shows on the lawn of Hildred's Hotel. In 1906 the same idea of afternoons on the sands and evenings inland was extended to Mablethorpe. Here the stage for concert parties (rather than pierrots) managed by Jack Clements was for many years on the beach just south of the pullover and later on the north side of the outfall promontory (taken over in the 1930s by Arthur Sherwin's Gay Cadets). Evening shows, and during the day if it was wet, were in the Victoria Pavilion Theatre (later a cinema and bombed in 1943). Some members of the concert party at its height in the 1920s were comic Jack Morris, baritone Fred Clifford, ventriloquist Joe Cookson, light comedian Len Sullivan, pianist Arthur Rawson, Ronnie Taylor the song and dance man and Lucy Clements (Jack's wife). The men wore blazers of a different colour each year. The pierrot show at Sutton, near the lifeboat station, was never a success and lasted only a short time.

1912 was a significant year for Fred Clements in Skegness. He opened the Arcadia Theatre on Drummond Road for evening shows with Harry Thorley's Band, and showed electric pictures at the Lawn, using gas generators. When the concession for the picture house passed to De Monds (who had their own Pavilion Pierrots), Fred built the Tower Picture House, designed as a real theatre complete with dressing-rooms. It was from there that much of the preparation was undertaken for his Christmas pantomimes in Leicester, Sheffield, Cheltenham, Plymouth, Worcester and Reading. For three summers after 1918 Jack Clements managed a show on the pier as well. Stars of the Clements Royal Entertainers of the 1920s were comedians Jimmy Loft, Charlie Harrison, Rob Currie and Ted Cartwright, pianist Gordon Webster, and Mabel Hind (Mrs Fred Clements) with saucy songs like 'Oh, Charlie take it away'. In 1932-33 Mai Jones was one of the pianists and in 1934 Avril Angers was a dancer when Kay Kendall went to watch her father in the show; Elizabeth Allen first had the urge to go on the stage when watching the Arcadia shows. Fred Clements, Skegness showman extraordinary, sold the Arcadia Theatre in 1935. He died in 1941 and appropriately was buried in the churchyard of St Clement's.

The new pavilion.

ABOVE: The interior of the new pavilion on Cleethorpes pier. CENTRE: Pier head cafe built in place of burnt-out pavilion. BELOW: A stroll on Cleethorpes pier.

ABOVE: The stump of the pier today. (LL) BELOW: Skegness pier under construction.

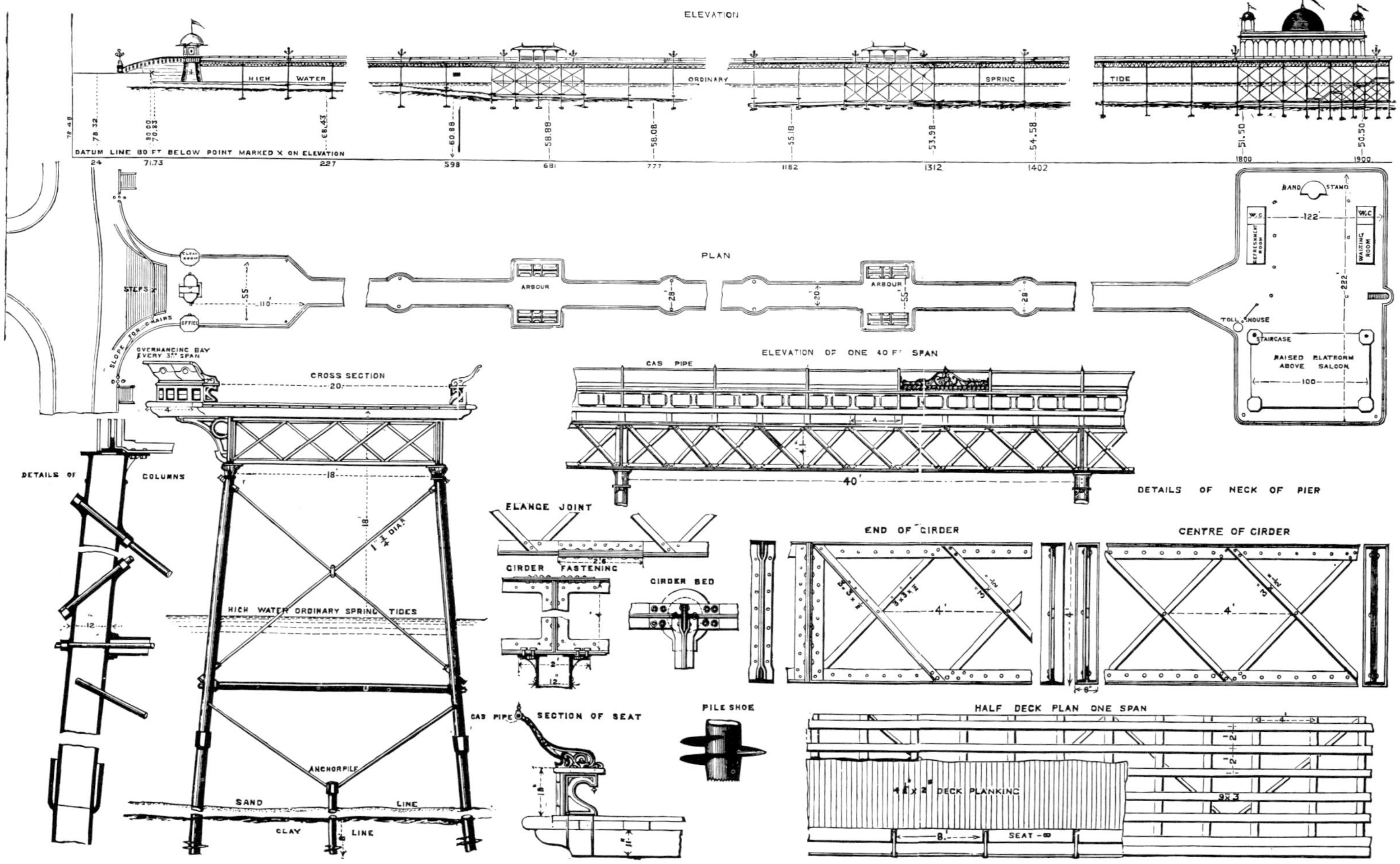

Iron Promenade Pier, Skegness 1880. (E)

ABOVE: The *Europa* which crashed through the pier in March 1919.
BELOW: Bridging the gap in the pier, 1919.

ABOVE: Pier entrance 1927. CENTRE: SS *Privateer.* (WK) BELOW: New pier entrance, 1938.

ABOVE: The tangled remains on the morning of 12 January 1978. (GET) CENTRE: All that remained of Britain's 4th longest pier. (JS) BELOW: Pierrots at Sutton-on-Sea—a Nainby photograph. (WERH)

ABOVE LEFT: Fred Clements, CENTRE: Jack Morris, and RIGHT: Jack Clements. LEFT: Jimmy Loft. (WK) RIGHT: Ventriloquist Joe Cookson.

ABOVE: Deck chair entertainment from a Clements star. CENTRE: Fred Clements Entertainers, Skegness. BELOW: Burrows and Chilvers Variety Entertainments—in *The Jungle* at Skegness, 1920.

ABOVE: Clements Arcadia. BELOW: Interior of the Arcadia Theatre (built 1910-1911 for £1,573).

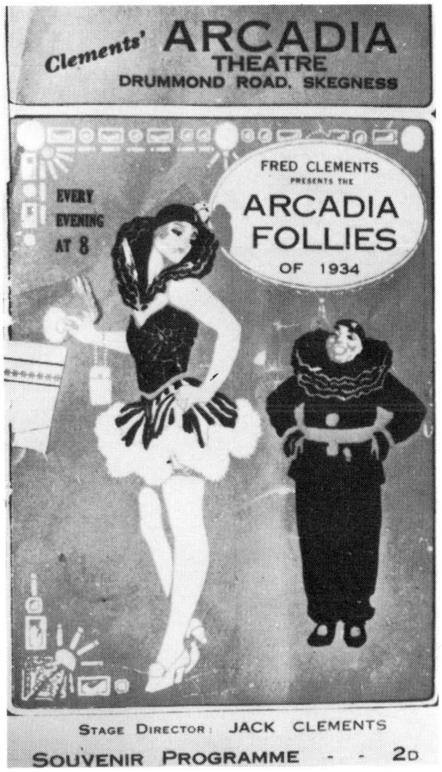

LEFT: Jimmy Jay and Bertie Rowley—entertainers at Sutton-on-Sea, 1908. RIGHT: Arcadia Follies, 1934. BELOW: Miss Nellie Dixon's Jolly Mascots, Cleethorpes 1905.

ABOVE: Clements stage and audience on the North Beach Mablethorpe.
BELOW: The Rollicking Recruits, Skegness 1914.

Launch the Lifeboat

The wind was north-east force 7, with good visibility despite intermittent sleet and rain, when part-time Coastguard Ivan Cartwright first saw the *Anzio 1* south-east from the look-out post at Donna Nook. It was 11.25pm on Saturday 2 April 1966. The vessel was about a mile and a quarter from the shore and travelling north. The coastguard signalled 'U' with the Aldis lamp—'You're too close to the shore'. But within a minute or two he could see both port and starboard lights as well as the masthead light, then port only, followed by both lights again, as if the ship had run aground. Immediately he informed his superiors and fired the maroons to alert the rescue teams. As he was speaking to Spurn Head to call out the Humber lifeboat, a red parachute flare was fired from the ship, and he fired two rockets to acknowledge the signal. It was 11.33pm. A few minutes later there was a second flare from *Anzio 1* and she could be seen on a sandbank a short distance from the shore. But low water had been at 10pm and the tide had still some 13 feet to rise.

When the first flare was fired Alan Loughton was checking poultry behind the sea bank. Sending his cousin for assistance, he ran across a mile of sand, mudflats and a water-filled runnel. The ship had a heavy list to starboard, away from and parallel to the shore. Wading out as far as he could, into about four feet of water, his shouts were answered by five figures. They were only about thirty yards away, but did not jump. The tide was coming in quickly and there was nothing more Alan Loughton could do.

Meanwhile a rescue team had been assembled by Coastguard Station Officer William Craigie, and their land-rover reached the water's edge about 12.42am. By that time the sea had advanced some 500 yards, too great a distance for the rocket lines. Eight minutes later Spurn lifeboat arrived, having been launched at 11.50pm and struggled the 11½ miles through 12-15 feet waves in a force 9 wind. Despite the conditions and a 4-knot tide running, the coxswain managed to get the lifeboat within 30 yards of *Anzio 1*. Sweeps of the searchlight revealed no sign of life, and instructions were given to search for survivors along the coast southwards, the direction of the set of the tide. The search continued until 8.35am, after a helicopter arrived at first light. The lifeboat arrived back at Spurn at 9.55am with two crew injured due to the rough seas. The *Anzio 1*, of 216 tons and launched in 1909, had been on the Scottish West Highland service before becoming a passenger vessel in the Thames estuary. She was bought by Cromarty Cruises Ltd and, after a refit, was making the voyage back to Scotland. As the weather had worsened, Captain Adam Fotheringham tried to turn into the Humber estuary for shelter—but failed. All 13 men on board perished.

Anzio 1 was not the first ship to be wrecked on those wide sandy shores which guard the south side of the entrance to the Humber. The broad promontory of the Sand Haile, and the shifting shoals and mud banks of the estuary have long been a hazard to shipping. The buoying of navigable channels was undertaken by the Brethren of Hull Trinity House until the formation of the Humber Conservancy Board in 1908 (and now, since 1968, the responsibility of the British Transport Docks Board). It was the Brethren who required trees in front of

Killingholme church to be cut down in 1828 so that it was clearly seen as a mariner's landmark, and the church tower whitewashed seven years later. Lighthouses are a rarity on the Lincolnshire coast, but the Brethren built two of brick at Killingholme—the High Light and the Low Light—for £350 in 1836. Sailing up the Humber, when the two lights 'were in one they will bear NW ¼ N and will lead ships and vessels . . . a clear course . . . from Spurn'. The lighthouses were later raised in height to 60 feet and 46 feet. Land beacons were also established by Trinity House, 64 feet, octagonal and black at Clee Ness in 1834 and 60 feet with triangular topmark at Donna Nook the following year. The one at Clee Ness was removed in 1863, but a new beacon had been built at Donna Nook in 1852, and in 1886 the first automatic bell buoy in the Humber was moored on the Sand Haile. Despite all these aids, in the 1870s there was an annual average of ten wrecks, by collision or grounding.

The causes of most wrecks were adverse weather and sea. With much coastwise traffic and many small boats plying in and out of the Humber and the Wash, the Lincolnshire coast had its share of wrecks. During the NE gales of 31 August and 1 September 1833, between 30 and 40 vessels were wrecked or grounded. Three Boston boats—the *Hope* with oil cake, the *Mary Ann* with wheat and the *Lively Parr*—together with a pilot boat and fishing cobble were driven on to the Holbeach marshes; the *Swift* of London, in ballast, was on shore near Gibraltar Point and three boats, including the *Hildon* of Whitby, were wrecked at Mumby Chapel and Huttoft; on the Humber mudflats were eight Cleethorpes fishing smacks and the sloop *Hopewell* from Thorne which carried a cargo of cinders for Louth. At Skegness the crew of the *Harmony* of London, laden with coal, were saved by the Skegness lifeboat although the next day the shore was 'completely strewed with dead bodies', clothing, coal and timber from that and other wrecks. (Many farm buildings on the coast were built of timbers from wrecked ships, collected as flotsam, removed from the wreck itself, or perhaps bought at wreck sales such as that of July 1876 when 5,000 feet of oak and elm planks and 500 oak posts removed from the 200 ton schooner *Naomi* were auctioned at Chapel.) The notorious Sand Haile or Hale claimed six victims including the *Amity* with china clay from Cornwall, the *Wear* of Sunderland, the *Henderika* with coals from Sunderland to Amsterdam, and the *Hermione* also with Sunderland coal. Farmer Richard Hoodless of North Somercotes was awarded the silver medal of the Royal Humane Society for swimming his horse through stormy seas to the wreck of the *Hermione* and saving four lives; the Donna Nook lifeboat could not get to it.

The first lifeboat on this northern part of the coast was established at Saltfleet in 1828 by the Lincolnshire Coast Shipwreck Association (of which Arctic explorer Sir John Franklin was Vice-Preident for many years). The idea of the Association originated at the end of the 18th century in a proposal from Sir Joseph Banks for a subscription for a lifeboat to aid William Marshall of Theddlethorpe Hall (where he had a 43-ft observation tower), the agent of London underwriters who had provided him with a sloop (with which he had saved 22 vessels in three years). In 1829 the Saltfleet boat was moved to Donna Nook, and Manby's Mortar apparatus for firing a lifeline stationed with the lifeboat at Theddlethorpe. Lack of funds prevented the Association from stationing another boat at Grainthorpe Haven near Sand Haile, but there was to be a lifeboat at Donna Nook until 1931. During that time there were 86 launchings and 172 lives saved, including 23 men from launches in 1864, 1865 and 1867. Two years later, the week of 20-27 October was stormy: 31 lives were saved from the steam whaler *Diana* of Hull and 14 from the barque *Bartolomeo Cerrutti* of Genoa, and 40 men from six brigs and schooners were assisted to land by local people—including farmer William Robinson who received a parchment from the RNLI (which had taken over administration of all Lincolnshire lifeboats in 1864). The Dobsons, a local family with long connections with the lifeboat, included Joseph, who rescued the crew of four from the brig *Amos* of Stockton in his own boat on 8 May 1843 and James who, with William Risdale, on 7 March 1844, rescued on horses the master and mate of the *William Septon* of Goole. On 27 May 1894 the lifeboat landed the master and 17 crew of the steamer *Arthur*. This lifeboat was the first of three to be called *Richard*, all presented (this one in 1887) by Miss Ann Dixon of Holton-le-Moor in memory of her brother. On Christmas Eve 1921 *Richard III* capsized in tremendous seas after rescuing four men from the Danish fishing boat *Koivisto*; eleven of the crew and the four Danes were flung into the sea, but all survived when the lifeboat righted itself. The thanks of the Institution inscribed on vellum, plus a double monetary award, were presented to Acting Coxswain John Dobson and the twelve members of the crew. It should be remembered that not only was the lifeboat still rowed (with ten oarsmen), but that all except the coxswain were farm workers. Also, before the provision of a motor tractor (in 1922, made by Claytons of Lincoln, modified by F. A. Standen of St Ives, cost £460) the boat had to be hauled a mile or more across the beach by a team of horses.

Of other lifeboats on the Lincolnshire coast, there was one at Cleethorpes from 1868 to 1882. This was the *Manchester Unity* under coxswain Levi Stephenson; when it arrived on 17 August 1868, with bands and under triumphal arches, some 50,000 people lined the cliffs and sands. Because of launching difficulties it was moved to Grimsby where there were two boats for a time (from 1898 to 1903 one was steam powered); the last was the *Thomas Burton*, when the station closed in 1927. One member of that crew was a Perrin who had lost his son and two grandsons when their fishing smack capsized at Saltfleet in 1912, and who was related to the Perrins of the Skegness lifeboat. There were lifeboats at Mablethorpe from 1863 to 1920. The *Heywood* was presented by Heywood Lonsdale in 1883 at a cost of £70 including the carriage, and the *John Rowson Lingard* was christened in 1905 by Baroness von Eckardstein. The eight horses were provided by Joseph Bell (a descendant of Ned the smuggler, who was crewman for 16 years and 27 years as signalman), Stephen Gray and Mablethorpe Hall Farm. The main wheels of the carriage were fitted with Tipping's plates or broad sectional tracks, as were other Lincolnshire stations after the 1880s, to prevent sinking in soft sand. The horses were ridden into the water until they were almost floating and then turned round to pull the carriage out as the lifeboat slipped off. When the boat had been rowed far enough from the shore, the sail could be hoisted.

For two years there was a lifeboat at Huttoft before it was moved to Sutton (on Sea) in 1844. After it became a RNLI station in 1864 there were four boats—*Birmingham No 1, Carolina, Elizabeth Berry* and *Sir John*—until it closed in 1913 with a record of 23 launches and 28 lives saved. There had been a lifeboat at Theddlethorpe from 1829, which continued under the RNLI until 1882. Further south the station at Chapel (St Leonards) was opened in 1879 with *Godsend*, followed by the *Landseer* and *John Alexander Berry*, the latter christened in 1888, the same year as *Elizabeth Berry* at Sutton. When the station closed in 1898 the record was 15 launches and 14 lives saved. Apart from practice launches, which also helped to raise funds, the Lincolnshire Coast Shipwreck Association held an annual rowing regatta for the lifeboats. The September day for the Skegness Life Boat Regatta in 1845 was 'beautifully fine and . . . several thousands of persons, comprehending almost every grade of society, were induced to assemble . . . some for pleasure, and some for the nobler purpose of patronising the Wainfleet Branch of the Shipwrecked Fisherman's and Mariners' Society' to watch a race rowed round a cutter stationed three-quarters of a mile offshore. Skegness won in a time of 16 minutes. In 1853 the race was also won, against opposition from Sutton, Theddlethorpe and two boats from Donna Nook, by the Skegness lifeboat under coxswain Samuel Moody.

Skegness is now the only Lincolnshire station with a large lifeboat. It is the *Charles Fred Grantham*, a 37-feet Oakley self-righter named after the farmer who was honorary secretary of the lifeboat station for forty years (1882-1922) and who provided the horses for launching. It cost £33,000 in 1964 and the launching equipment a further £12,000. There is a crew of six under coxswain Ken Holland who has been a lifeboatman since 1946. The first boat cost £130, had no name, was 30 feet long, had a crew of ten (eight of them oarsmen) and was steered with a sweep oar. It was established at Gibraltar Point in 1825, two years before the formation of the Lincolnshire Coast Shipwreck Association, under coxswain William Scupholm, who was licensee of the Old Ship Inn. Despite the advantage of Wainfleet Haven, Gibraltar Point was too far to the leeward of sandbanks likely to trap vessels running for shelter in the Wash, and in 1830 the lifeboat was moved to Skegness and housed in the dunes where is now Lifeboat Avenue (and later, in 1864, to the present site on South Parade). It also marked the start of 41 years' service as coxswain by Samuel Moody; he was awarded two silver medals—for saving 53 lives in 21 years, and for the rescue of 11 people (including the captain's wife and child) from the burning timber-laden brig *Atalanta*, when it was driven onshore three miles north of Skegness, during a north-easterly gale on 18 October 1854 (a night when 22 vessels were wrecked on the Lincolnshire coast).

One of the problems of launching a lifeboat on the shelving sands of the Lincolnshire coast was that of having to row it through strong surf, more often than not into the teeth of a gale. The Skegness lifeboat was twice driven back by heavy surf on 3 January 1830, when it was launched to the assistance of the sloop *Thomas and Martha*, aground on the Skegness Middle sandbank two miles offshore; on 2 December 1867 the boat was pulled three miles along the beach (there were no groynes to impede passage then) to get within launching distance of the brig *Claudia* of Sunderland which had gone aground at low water at Ingoldmells; on the night of 5 December 1875 the lifeboat again had to be hauled two miles along the shore in snow showers before it could be launched through the surf into an easterly gale, to get to the *Star* of Colchester ashore at Winthorpe Gap (a night when Coastguard George Chesnutt and another Samuel Moody were awarded silver medals for rescuing the captain, all three having to be pulled ashore clinging on to outside lines); on 26 February 1916, it took 80 soldiers with ropes from the pier to pull the lifeboat clear of the breakers, in order that survivors could be taken off the 2,130 ton steamship *Bogatyr*, carrying iron ore from Bilbao to Middlesbrough, stranded on sandbanks off Chapel pullover; and on the last service launch using horses on 22 October 1924, the lifeboat was nearly driven on to the pier by the wind and surf.

Those last two launches were in the period when the *Samuel Lewis*, a yawl-rigged 35-feet Liverpool-type lifeboat presented by Mrs Ada Lewis-Hill, was in service (from 1906 to 1932, with a motor tractor used for launches from 1926). It was also when Matthew Grunnill was coxswain (1908-1932), one of a family which has served Lincolnshire lifeboats for 150 years. There were eight Grunnills in the crew of 17 in the Skegness lifeboat, in 1912. Son of the coxswain of the Chapel lifeboat, Matthew started at the age of 16 and served in all three boats there. He was an outstanding coxswain at Skegness despite spinal injuries, suffered when he fell almost 40 feet from the flagstaff crosstrees when in his late thirties. His cousin Montague, who served a total of 41 years from 1893, was also appointed 2nd coxswain in 1908. There were six Grunnills and four Moodys (another famous Skegness lifeboat family which produced two other coxswains besides Samuel—Joseph, 1877-80, and John Smith, 1900-08) in the boat on the occasion when it put out on 19 November 1893. The previous day one of the worst gales remembered had done much damage to buildings in Skegness and the beach was littered with the wreckage of ships. It took a quarter of an hour to row through the surf before the sail could be hoisted to chase after the Grimsby cod-smack *Frank* which, with its mast and rigging blown away, was rapidly driven into the Wash and towards the Norfolk coast. The crew of seven were taken off and landed at Sutton Bridge on the River Nene, but it was three days before the lifeboat got back to Skegness.

In those same gales Montague lost his brother Dashwood off the Norfolk coast, and earlier in the year his father and another member of the Grunnill family (Edward and Edwin) had been drowned in the worst disaster off the coast. On 8 July the pleasure sailing yawl *Shannon* capsized in a dramatically sudden squall off Skegness before the sail could be hauled down, and 26 people were drowned. Three survivors were picked up only because another Grunnill, Jabez, was fishing nearby in his own boat the *Dart*. No wonder that in his lines on the tragedy the blind Grimsby poet C. J. Parker wrote, 'These Grunnills had great praise for their kind and manly ways'.

Matthew and Montague Grunnill were themselves saved in 1905, when fishing for cod. The lifeboat then was the *Ann, John and Mary* (1888-1906), the only other self-righter to be stationed at Skegness, and to speed the launch a large crowd manhandled the boat to the water's edge. The previous boats had been the *Herbert Ingram I* (1864-74) and *Herbert Ingram II* (1874-88) presented by Mrs Ingram in memory of her husband, a native of Boston who founded *The Illustrated London News* and was himself drowned in Lake Michigan in 1860. For their part in one of the Skegness lifeboat's most famous rescues, Matthew and Montague were given silver medals by King Haakon of Norway, with £20 divided between the rest of the crew. On 13 November 1912 they rescued eight men from the 300 ton Norwegian brig *Azha* aground on Skegness Middle, with a waterlogged cargo of pitprops and all sails stripped off by the gale.

The third family closely associated with the Skegness lifeboat is that of Perrin. George was one of 16 children born to James, harbour-master of Wainfleet Haven, who lived in Noah's Ark at Gibraltar Point. This was a fishing smack grounded on the marsh about 1870 and converted into two rooms, where the summer visitor could buy ginger beer, lemonade and nectar. George joined the lifeboat crew in 1912 and became coxswain in 1932 when the first motor-driven boat came into service—the *Anne Allen* (35½ feet, single crew, with a range of 50-65 miles and a crew of eight). In May 1935 the lifeboat was called out to the disabled tramp steamer *Creek Fisher* near the East Dudgeon lightship, 40 miles out into the North Sea; the successful navigation of the dangerous shoals off the north Norfolk coast was a demonstration of George Perrin's fine seamanship learnt in smacks and trawlers, most of it off the Lincolnshire coast and in the Wash. Two months later when the lifeboat was undergoing a refit, George and his son Wilfred (later coxswain from 1947 to 1965) along with Percy and Lance Grunnill, ex-coxswain Matthew Grunnill (then in his late 70s) and James Giles, manned the beach pleasure boat *Grace Darling II* (the former Coverack lifeboat) to rescue two men in a dinghy which had been blown out to sea.

The 148 ton steel sailing barge *Britannic*, which had been rendered assistance by the *Samuel Lewis* in October 1912, called out the *Anne Allen* in November 1937 when she lost her rudder and was driven into the Wash by an easterly gale; the crew of three men and a dog were saved. One of the quickest launches of a Skegness lifeboat was on 23 November 1938 when it was in the water only 14 minutes after the maroons were fired, despite the fact that it was low water and almost impossible to stand upright on the beach in a south-westerly gale gusting up to 70 mph. On that occasion three men were saved from the Boston drifter *Dusty Miller* which sank soon afterwards, the name board eventually washing up on Skegness beach. During World War II many launches were made, mainly to search for crews of ditched aircraft, including one occasion in October 1940 when it took some plain speaking by George Perrin to get the Army to open up the barbed-wire barricades on the beach. A regular run for many years for the lifeboat was to take Christmas fare to the Lynn Well lightship in the entrance to the Wash, until it was replaced by a light buoy in 1973. In post-war years the main problems have been with air beds and rubber dinghies out to sea, or engine failure on pleasure boats. This happened to a 16-year old lad in a 12-feet speedboat only half a mile off Anderby Creek on 17 June 1961; Wilfred Perrin and lifeboat *Cuttle* (1953-64) failed to find him, and it was the next day before he was spotted by a Swedish steamship—21 miles away.

There may no longer be wreckers on the Lincolnshire coast, if there ever were, or opportunities like that presented to Cleethorpes when the SS *London*, with a cargo including red flannelette, geese and ducks caught fire and ran aground one December. Many of the birds finished on the Christmas table, and the material made warm undergarments for most of the population—to become the mark of a genuine 'Meggie' (and fragments are still known to be carried in wallets and handbags). In other places the name of a wrecked ship lives on, like *Rimac* House and the *Prussian Queen* pub at Saltfleetby. (*Rimac* derives from the name of the 215-ton Blyth brig wrecked near Saltfleet on a voyage from Cronstadt to Hull on 11 December 1874.) But there will always be a job for the coastguard, though perhaps no longer at Gibraltar Point, where a station was built in 1859 on a narrow neck of dunes between Wainfleet Haven and the open foreshore and permanently manned with its own boat until 1922 (by which time it had been separated from the sea by a new dune and marsh growth), while certainly at Skegness, Mablethorpe and that loneliest of all look-out posts on the Lincolnshire coast—where a three-year posting is the maximum for any Coastguard Officer—Donna Nook.

ABOVE: The wreck of the *Anzio*, April 1966. (DD) INSET: Clee Ness (left) and Donna Nook (right) beacons, 1835. BELOW: The ribs of a wreck on the beach between Pye's Hall and Donna Nook.

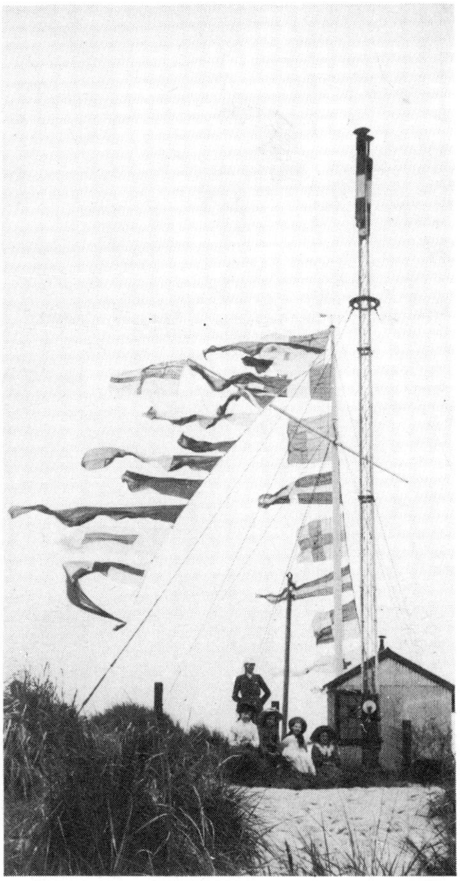

LEFT: One of a pair of lighthouses at the entrance to the river Nene; once occupied by Peter Scott. (LL) RIGHT: Coastguard Station, Mablethorpe 1908. CENTRE: The Grimsby lifeboat and crew, c1904. BELOW: Painting of the Donna Nook lifeboat *Richard* 1878.

ABOVE: The Donna Nook lifeboat *Richard,* c1900, and RIGHT: Captain Dobson. CENTRE: The *Heywood* lifeboat at Mablethorpe—presented in 1883, (RWD) and BELOW: Mablethorpe welcomes Baroness von Eckardstein who christened the *John Rowson Lingard* in 1905.

ABOVE: The *Sir John* lifeboat at Sutton-on-Sea—a Nainby photograph. (WERH) CENTRE: Old Lifeboat House Skegness 1870. BELOW: *Ann, John and Mary*, Skegness, c1904.

LEFT: The *Samuel Lewis* (1906-32) under sail. (RBW) RIGHT: The present Skegness lifeboat *Charles Fred Grantham*. CENTRE: *Samuel Lewis* being towed by the first tractor on service at Skegness 1926; driver Wilfred Grunnill. (WK) BELOW: Crowds watching a lifeboat launch at Skegness c1905.

The Great Storm Surge of 1953

TIDAL TORRENT SMASHES SEA WALLS
THROUGH THE DOOR, A SURGE OF WATER
THE SEA TAKES 20,000 ACRES
ARMY TAKES OVER THE 'LINCOLNSHIRE DUNKIRK'

Newspaper headlines spelt out the stark facts of the few short hours from 5.25pm on Saturday 31 January 1953 which claimed 41 lives. The human consequences of the flood, including the total evacuation of Mablethorpe and Sutton on Sea, were greater than any previous catastrophe of similar magnitude because of the growth of holiday resorts and retirement homes. It was the greatest disaster the Lincolnshire seaside has ever known. How did it happen?

The brutal fact is that the motley collection of sea defences—sandhills, clay banks, wooden walls, poor quality war-time concrete, slabs of concrete on sand and isolated steel-piled stepped concrete walls—were inadequate to withstand the onslaught of mountainous waves driven by near-hurricane force winds. The meteorological factor was crucial. Two days before the disaster a low pressure area forming to the south-east of Iceland was christened Depression Z. This moved, first north-east and then south-east into the North Sea, deepening to a pressure of only 966 millibars by mid-day on Saturday 31st. At the same time a ridge of high pressure built up to the west of the British Isles, creating an unusually steep pressure gradient. The result was abnormally strong winds blowing totally unimpeded from the Arctic into the North Sea towards the exposed coast of Lincolnshire. The effect was threefold. The combination of low atmospheric pressure and high winds raised the moderate full moon high spring tide seven to eight feet above the predicted level on the Lincolnshire coast, that is some ten feet above the lowlying lands of the Outmarsh. Secondly there was a progressive piling-up of wind-impelled water, as it was deflected along the east coast of England and moving south—the surge. Fortunately the surge did not coincide either with the highest spring tide of the period (at the new moon) nor yet with the actual time of high tide. Only as the tide was falling and the evening hours wore on did the surge arrive. The effect was to prolong the period of high tide. Then there were the waves—high (at least 20 feet) and steep, the most destructive type. They were generated by the very long fetch, the length of sea track over which the wind blows, and the strength of the wind. Moreover the wave effect was increased by the very high water level.

The gales which sprang up on the morning of that fateful Saturday and continued into the afternoon did not give rise to any particular alarm. The Lincolnshire River Board instituted only the normal precautionary measures, placing storm patrols on duty. Mrs D. M. Morley of High Street in Sutton on Sea put the finishing touches to the baby's christening cake ready for the next day, and calmly thought that the evening tide would be a sight worth watching. Mrs A. V. G. Critchley of Sutton went out to buy fish for the evening meal; her husband had been out of hospital only a week; they lived within 200 yards of the sea wall. In Mablethorpe PC F. Midgley was on his beat and newspaper reporter Elizabeth Chesterton was out collecting local stories. Many others, like 80-year old Mrs Annie Millward who lived on her own at Saltfleet, Mr and

Mrs T. Walker at Chapel St Leonards, and retired Baptist Minister Rev Walter Lait and his wife Amy at Ingoldmells, secured the windows and made sure there was plenty of coal inside for a warm fire that evening. Four Leicester people, Mrs Usher, daughter Mrs Thelma Broadbury with her brother-in-law Mr Clifford Harrington and his six-month baby, wondered what sort of winter holiday weekend it would turn out to be in Sutton on Sea. The whole Lincolnshire seaside was soon to find out.

As darkness fell, the wind had veered to north and blew with increasing force. As the tide rose, the pounding breakers moved closer and closer, their massive undertow scouring sand from the narrow beaches and then clawing it from the sandhills. Concrete sea walls were pounded and began to crumble. But still the tide continued to rise until the waves crashed against the dunes behind the sea walls or washed over the top and began to erode the landward slope of the defences. Completely unaware of the impending disaster, life took its normal course. At tea-time the water started to come over Sutton pullover. It had happened before. But the trickle became a steady flow and the Morleys started to take precautions; there was hardly time to do anything before there was a foot of water in the scullery. The message that waves had begun washing over the sea defences at Sutton was received by River Board Engineer F. H. Tomes at Boston about the same time as the River Witham overflowed into the streets there.

At 5.25pm the sea defences gave way completely at Sandilands. The sea blasted a quarter-mile breach in the timber-piled dunes—the notorious Acre Gap—spreading a huge fan of sand over the fields behind. There was nothing between the sea and the edge of the Middle Marsh two miles inland—and the tide flowed unimpeded through the gap. For the next two hours the sea pounded over ten more major breaches in the defences between Mablethorpe and Ingoldmells, and wrecked the promenades at Mablethorpe (where blocks of the stepped concrete merely rested on baulks of clay with sand between) and Sutton (where part of the late 19th century simple vertical wall was flattened). One of the key breaches in the promenade at Mablethorpe was behind the Council offices. This gave way at 7.10pm and the sea, as at Bohemia near Trusthorpe and at Sutton, drove masses of sand and water into the streets. An estimated 860,000 tons of sand were washed into Mablethorpe, and there was up to eight feet of sand in the streets and round the houses of Sutton.

It was 6.15pm when Mrs Critchley went to see how the fish was cooking in the oven and found water coming under the door, then through the french windows and up through the floorboards. As they retreated upstairs the lights fused. By 7.30 the water was five feet deep, with a wave-like motion which smashed furniture against the walls. Many others had a similar experience. Mrs Morley scrambled to get baby, baby food and dry clothing upstairs (not forgetting the christening cake); from the upper windows the main street was a racing stream of flotsam.

Some were caught out of doors. John Manning, struggling to reach his home in Somersby Avenue, Mablethorpe, was caught in rapidly rising water up to his chest. Swept off his feet, he managed to clutch the gate and haul himself on to the coping of a wall where he remained for two hours until the water level fell with the receding tide. Newspaper boy Malcolm Hill out on his round was rescued by PC Midgley who saved sixteen other people, some from bungalows, and sheltered them in his own home. When Mrs R. Wileyman opened the front door to put the Sunday newspaper money under the mat, a wave of water surged into the house. When Elizabeth Chesterton reached home in Tennyson Road, Mablethorpe, at 7 o'clock there was no sign of floods, but an hour later the water burst through the front door and put out the fire. Quickly getting everything possible upstairs, the family lit a fire in the bedroom fireplace to keep warm—burning seven pairs of slippers, two lamp standards, coat hangers and a fibre suitcase. The family holiday party at Sutton all drowned, despite the valiant effort of Alex Henshaw (winner of the 1938 King's Cup air race and wartime Spitfire test pilot) who rescued Mrs Thelma Broadway and took her to another house; she died shortly afterwards. One

Mablethorpe family, Mr and Mrs H. Lynne and their son, were able to escape the flood waters when they found a hand-operated plate-layer's bogie on the railway line.

Due to the alignment of the coast, Mablethorpe and Sutton bore the brunt of the attack. North of Mablethorpe towards Donna Nook the beach widens, itself a protection, backed by dunes and reclamation banks. But even here there was no escape, with many minor breaches in the banks and one major breach—at Sea Lane, Saltfleet. Only three months retired from Rotherham, Frederick Drury heard a roaring sound and saw a wall of water sweeping over and through the dune-covered bank. Nearby four elderly people drowned, including Mrs Annie Millward. She had arranged to ring a bell connected to the house next door if ever she needed help, but the neighbour herself was marooned and had to listen to the tinkling of the bell as it grew fainter.

As the tide turned into the Humber from Donna Nook, it was constricted by the funnel-shaped estuary and, despite the wide beaches and protection afforded by Holderness and Spurn Point, nowhere escaped the rising waters. Banks were overtopped, thousands of acres of reclaimed lands were inundated and RAF North Coates was evacuated. At Cleethorpes sand was washed into the bathing pool and on to the promenades, wrecking amusement facilities, particularly between the pier and Wonderland, which became a tangle of chunks of the promenade and timbers from the shattered Arcadia. Round the corner in notorious Suggitt's Bay, the sea blasted through the railway embankment, flooding nearly 1,000 houses. Holes were torn in the piers at the entrance to Grimsby docks. Between there and Killingholme Haven the Humber bank was reduced in height by some three feet and there were many minor breaches with flooding of factories; Immingham power station was put out of action, and two ships in dry dock were capsized.

Between Sandilands and Chapel Point the dune-covered earth seabank largely disappeared along many stretches, and there were major breaches at Sea Bank Farm, Elder Cottage and Chapel Six Marshes. Those who had ignored the Biblical injunction and built their holiday homes atop the dunes at Anderby Creek found the front rooms hanging precariously over the beach. However, the 'Roman' bank further inland, although weak, acted as an effective brake on the extending flood. Houses between the banks, as at Chapel Marsh, were totally wrecked, fortunately without loss of life. At Chapel St Leonards, Mr and Mrs T. Walker were alerted by their whimpering dog at 6.45pm and found the bungalow surrounded by water; they spent the night on high ground sheltering in bushes. The steel-piled defences of Chapel Point held, but the promenade and dunes behind were heavily battered. The main promontories of the Lincolnshire coast—Chapel Point, Vickers Point and Ingoldmells Point—originated from natural boulder clay hummocks or the need to protect drainage outfalls. Not only were their northern sides overtopped, but there was a sharp deflection of waves round them to increase the intensity of attack on the southern side as well. Where the latter coincided with dunes lacking in sand-binding vegetation due to constant passage of holidaymakers, as at Vickers Point, a breach was inevitable. Both here and at Trunch Lane near Chapel (with a massive 600-yard breach) which had also suffered from trampling, and where the dune line was narrow anyway, the empty caravan sites behind were reduced to heaps of unrecognisable jumble.

The patched-up stepwork north of Ingoldmells Point, known as 'Ingoldmells Mess', suffered badly and there were breaches at the Point itself. It was here that there was the greatest loss of life (17) because the houses and particularly bungalows were between the sea wall and the 'Roman' bank where water rapidly reached five or six feet in depth. Rev Walter Lait and his wife failed to get through a hole they had clawed in the bungalow ceiling, as did bus driver Bill Chambers, his wife, mother and three children. Mr J. Truman, deputy camp controller at Butlin's Holiday Camp (itself under six feet of water when the sandhills in front gave way) swam through the flood to help 13 people to temporary safety in a false roof.

Skegness escaped comparatively unscathed. Beach chalets were carried away, the well laid-out gardens disappeared under salt water for the first time in their seventy years' existence, the amusement park was flooded, there were four feet of water in the Sun Castle and the north-east corner of the pier was extensively damaged. The town was saved by the wide foreshore and the original promenade wall. The eastern dunes at Gibraltar Point were cut back to reveal war-time beach defences. As the surge swept into the dead-end of the Wash, the tide rose by more than six feet above the predicted level at Boston, sea banks (old and new) were breached or overtopped in the Welland and Nene estuaries,flooding fertile farmland.

When Mr Tomes and his deputy engineer were able to leave the flooded streets of Boston, they made for Sutton on Sea to be met by the tide flowing quickly inland. By a circuitous route to the north they entered Mablethorpe and scrambled across the dunes to make a torchlight inspection of the damage. Telephones and public services were out of action, so they returned without delay to set up temporary headquarters in the police station at Alford. Procedures outlined in instructions previously circulated by the Ministry of Agriculture and Fisheries were immediately put into action. Shortly after midnight Army and RAF personnel arrived with lorries and equipment. At 1am Elizabeth Chesterton and her family waded through waist-deep water to a rescuing lorry. Nearly 100 boats from Skegness boating lake undertook a new role; with one of them Mr Truman was able to ferry in twos and threes the 13 people trapped in the false roof at Ingoldmells.

The wind still howled as the lashing sea retreated across the sandless foreshore leaving death and destruction in its wake. It was a brilliant moonlit night. Looking inland towards the Wolds there was water as far as the Critchleys could see. The only other light came from the occasional candle in nearby houses. At 4am the Morleys were rescued—mother and children first, with the baby sleeping peacefully through it all and the christening gown left hanging high above the jumble of overturned furniture on the wet, slimy floor. The christening cake fed the hungry RAF rescuers. And so it went on throughout the night. At 7 o'clock the next morning came the next high tide and the water in the Critchleys' house deepened again to four feet. Daylight came and not a soul in sight; only the waves breaking over the lawn and garden. Then on the 9 o'clock news on the wireless they heard that Mablethorpe and Sutton were to be evacuated, but it was not until 2pm that the rubber dinghy sailed into the house and took them from the fourth step of the stairs.

By Sunday night most of the 5,000 population of the two towns had become refugees in Louth and Alford, where rest centres, canteens, food, blankets, clothing, billets in houses, transport, care and comfort were provided with astonishing speed. Meanwhile the coastline was invaded by engineers, contractors and thousands of Army and RAF servicemen. Thirty four miles of sea defences had been destroyed and breached, and each tide was flowing in and out of the flooded areas. They had a fortnight before the next high spring tides in which to prevent a greater disaster. It was a fortnight which had little sleep for Divisional Engineer John Finn who had arrived to take up his new duties at the Sutton office on 1 February. The emergency operations, dubbed 'the battle of the breaches', were a remarkable achievement. Under apalling weather conditions of rain, wind and snow 4,000 troops and workmen erected sandbag walls across the worst breaches (there was no shortage of sand), persevering to rebuild sections damaged by each tide. From the Humber to the Wash contractors were given instructions to bring in all the draglines, dumpers, piledrivers, bulldozers and other heavy equipment they could gather and to work day and night (using searchlights). They were each allocated a section of the most severely damaged defences. Two thousand lorries made two or three journeys daily, transporting Scunthorpe slag, chalk and other stone from as far afield as Derbyshire and Northamptonshire, pouring 25,000 tons of material into the breaches every 24 hours. The Acre Gap alone swallowed 40,000 tons of Scunthorpe slag, brought in by rail to Sutton, with a continuous line of lorries operating for four days. The test came on 16 February when the evening tide was 1ft 7ins higher than predicted (and only 1ft 3ins below that of 31 January); but there was a dead calm and all the works held.

On into the spring the contractors worked, rebuilding clay banks from Saltfleet to Cleethorpes, along the Humber bank and in the Wash, using material excavated from borrow pits in the fronting saltmarshes, repairing eroded dunes with brushwood kidding, and above all along the exposed holiday coast building new solid defences with interlocking steel piles driven firmly into the foreshore clays and concrete, or grouted stone walls topped by a curved wave wall. By the end of March the bill had reached £4½m, on top of which there would be the repair of the approach roads fractured by the abnormal weight and volume of traffic, at somewhere in the region of £150,000. Meanwhile hundereds of volunteers were clearing sand from houses, the RAF dried out furniture and carpets, and the fire services pumped millions of gallons of water back into the sea.

At the beginning of March the newspaper headlines spelt out a different message.

ON SUNDAY IT WAS OPERATION SCRUB
MABLETHORPE COMES BACK TO LIFE AND HOPE
SEASON 'AS USUAL' AHEAD

The sea defences were safe. The refugees returned. And Mrs Critchley finally got that fish out of the oven.

Shattered railings and debris strewn across the North Promenade, Cleethorpes. (GET)

ABOVE: Undermined and collapsed sea front at Mablethorpe—showing clay baulks which had supported concrete stepwork, but with sand between.
BELOW: Beach chalets at Mablethorpe perch precariously on a plinth of sand. (GET)

ABOVE: The major breach in the defences at Mablethorpe. BELOW: Air view of the same breach with construction of sandbag wall already started. (Note field boundary lines and diggings on foreshore clays stripped of sand.)

LEFT: No shortage for filling sandbags. (GET) RIGHT: Flattened sea defences just south of the Lincolnshire Convalescent Home. (GET) CENTRE: All that was left of the dunes at Trusthorpe. (GET) BELOW: One man and his machine against the sea. (GET)

ABOVE: All that was left of the promenade at Sutton (note the timbers of an earlier wreck exposed on the sandless foreshore). CENTRE: Vast quantities of sand were washed into the streets and houses of Mablethorpe and Sutton. (GET) LEFT: This horse did yeoman service where other vehicles could not reach. (GET) RIGHT: Evacuation. (GET)

ABOVE: The fan of sand swept in land from the notorious Acre Gap at Sandilands, and the flooded acres beyond. BELOW: The huge breach at Acre Gap.

ABOVE: A continuous shuttle of lorries deliver slag to seal off the Acre Gap. (GET) BELOW: Dune-top houses built at Anderby Creek between the wars left hanging over the beach. INSET: The Duke of Edinburgh visits devastated Sutton. On his right is John Finn, River Board Divisional Engineer who took up his post the morning after the flood. (GET)

ABOVE: Shattered remnants of seaside bungalows at Chapel St Leonards. Flood water behind has been contained by the 'Roman' bank. CENTRE: Waves breaking through the broken wall of Chapel Point on 1 February 1953. LEFT: A jumble of caravans at Ingoldmells. (GET) RIGHT: War-time defences revealed where the covering dunes were washed away at Gibraltar Point. BELOW: All that was left of the shelter on Ingoldmells Point. (Note the clay core of the sea bank exposed in the foreground.)

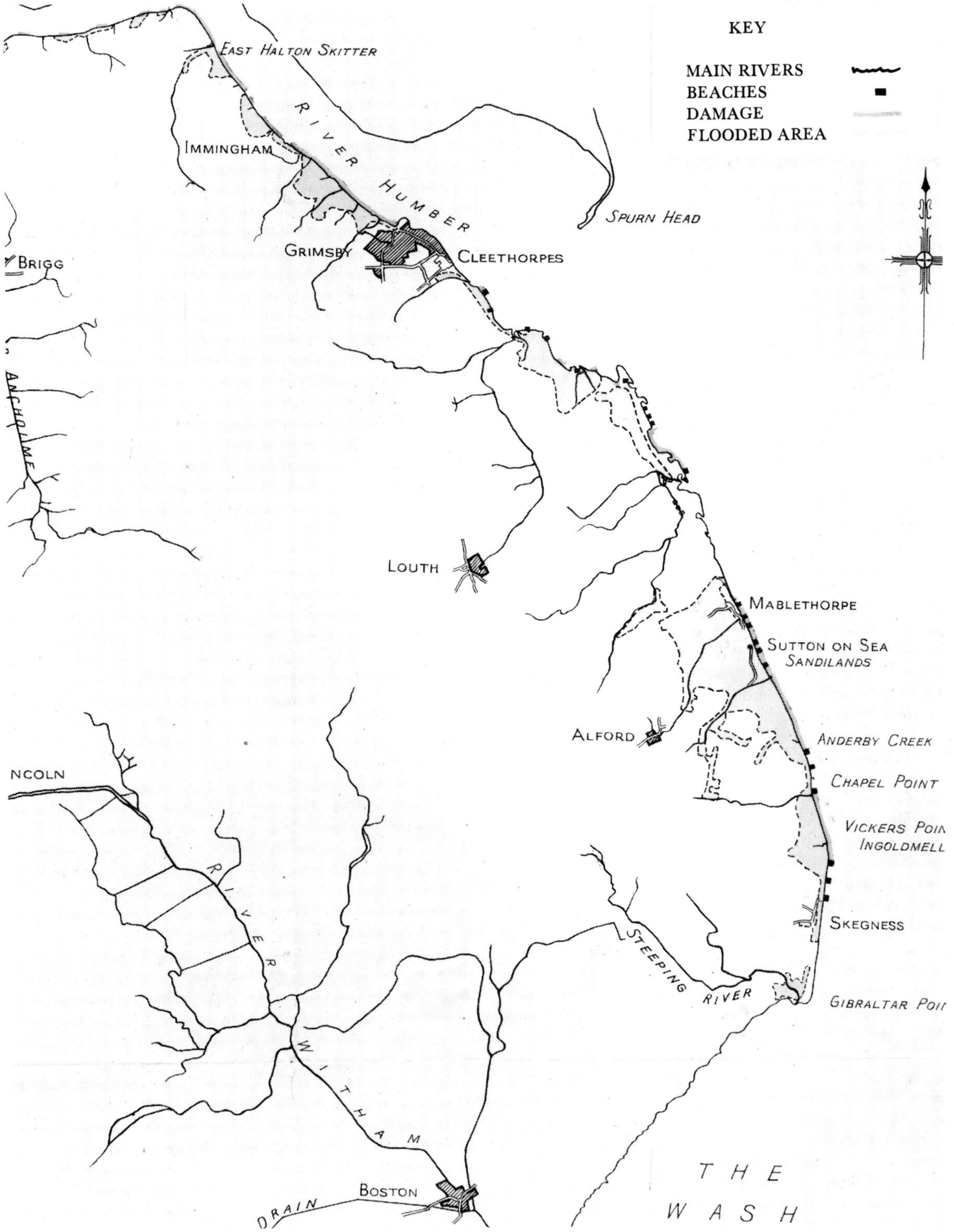

The extensive areas where the sea flooded inland.

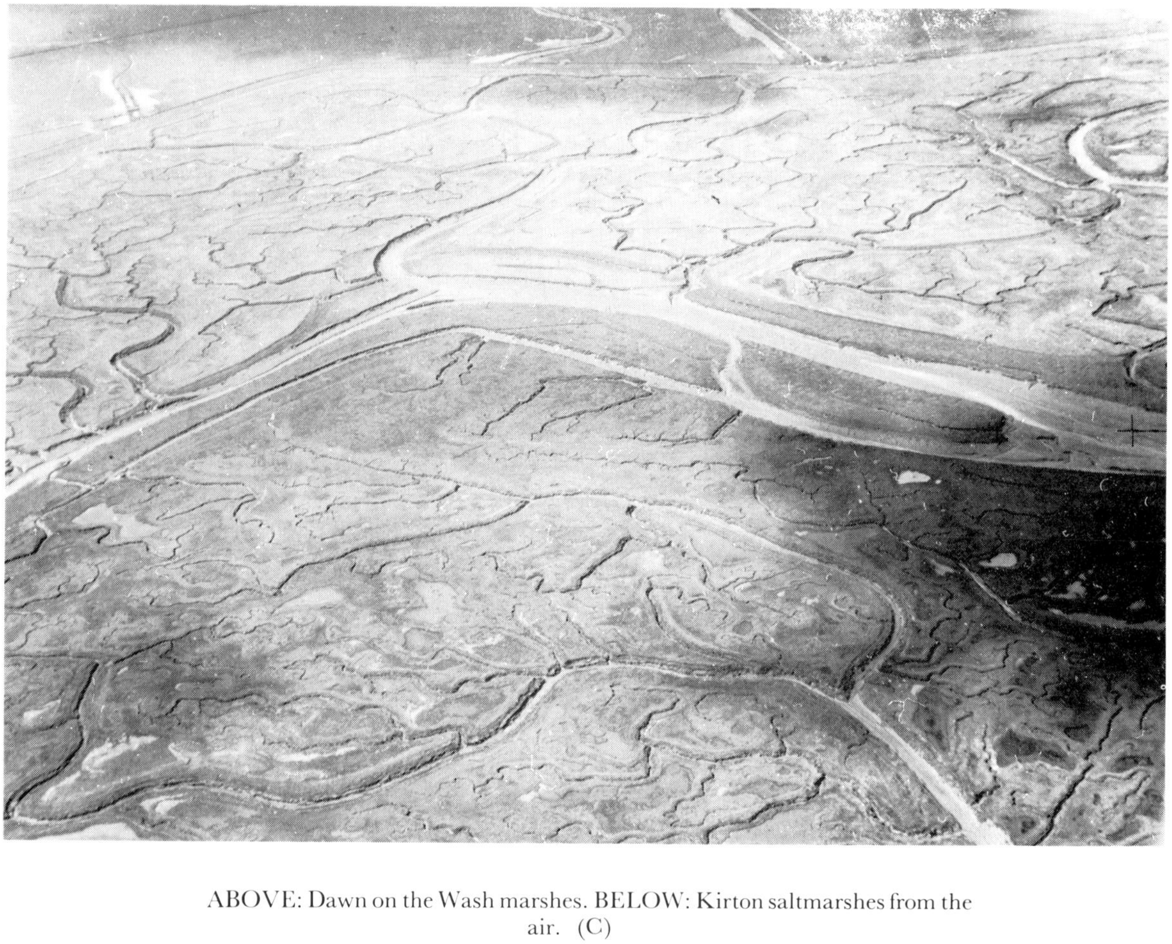

ABOVE: Dawn on the Wash marshes. BELOW: Kirton saltmarshes from the air. (C)

What shall we do with the Wash?

The Wash mud is clammy and cold
Here King John lost his riches of old—
    This may happen again
    It's increasingly plain,
Only this time it's your and my gold.

—D. C. Staveley

Ptolemy named it Metaris AEstuarium—a bay with tidal marshes and mud flats. Not until the 16th century was the name Wysche or Waashe recorded by the Commissioners of Sewers (land drains) but then only referring to the tidal inlets of the Welland and the Nene. Lincolnshire's own Fynes Moryson of Cadeby wrote around 1600 of 'the great Washes of Holland' which 'when the sea flowes are covered with water, but when it ebbes, the ground is discovered to be passed'. Nearly a century later Camden said of The Washes that they were 'a very large arme (of the German Ocean) and passing well knowne, at every tide and high sea covered all over with water, but when the sea ebbeth, and the tide is past, a man may passe over it as on dry land, but yet not without danger. Which King John learned with his losse. For whilst he journied this way . . . the waters suddenly brake in upon him, so that at Fossedyke and Welstream he lost all his carriage and princely furniture'. The exact location of that near fatal disaster on 11/12 October 1216 when King John misjudged the time of high tide on the Nene estuary has never been determined. It is reckoned to be somewhere between Walpole and Foul Anchor (near Tydd Gote), now nearly six miles from the sea.

Neither the King, nor yet Moryson or Camden could have foreseen the reclamations that would dramatically alter the shape of the Lincolnshire shores of the Wash. Even less could they have imagined that in the middle of the 20th century Fred White of Walpole would suggest throwing a barrage across the mouth of the Wash to create a massive lake, or that there would be suggestions for laying out a speedway, an airport or even raising a city on what Michael Drayton described in 1613 as 'the vast and quenchy soyle'.

Of the schemes that would have drained more money from the pockets of the investor and the taxpayer than water from the marshes and mud flats, the first was that of Nathaniel Kinderley in 1751. He was concerned with the 'pernicious sands', the continual shifting of which in the estuaries of the four Wash rivers impeded navigation. He proposed to unite the Nene with the Ouse, and the Welland by a new channel from Fossdyke to the Witham near Skirbeck Quarter, whence the united river would continue a straight course to the sea near Wrangle. The result of this, he considered, would be the entire silting up of the estuary and the gaining of 100,000 acres of 'new habitable country'. Across this he proposed a new road from Lynn to Boston. And the cost of the whole eccentric scheme?—a mere £150,000, surely a considerable underestimate even in those days.

A more serious and realistic proposal was that made by the great engineer Sir John Rennie in 1839. He recommended that the channels of the four rivers should be confined by fascine work (bundles of brushwood) and led to a common outfall near the centre of the Wash where such a confluent mass of fresh water in one channel would eliminate silting up. As the training walls extended, so the land between the outfalls would be embanked as it accreted. Some impressive supporting statistics were given: the gaining of 150,000 acres which in a few years would be worth £6m was to be set against estimated costs of only £2m. The meeting which received the report thought that as the cost exceeded the means of private individuals, the scheme 'ought to receive grave consideration and the eventual support of Her Majesty's Government, as a purely national object'. Such support was not forthcoming, and it was found impossible to raise the funds. The notion had been that the new land should be called Victoria County, and a map of about 1860 actually has the words printed over the mud flats, but no works ever started.

Eleven years later a part of the idea was revived in the Lincolnshire Estuary Company's Acts of 1851 and 1852 'for reclaiming from the sea certain lands abutting on the coast of Lincolnshire, within the parts of Holland'. The outer reclamation limits were much as those proposed by Rennie, except for a much larger area between the outfalls of the Witham and the Steeping River at Gibraltar Point—a line roughly coincident with the present low water mark of ordinary tides. To reclaim an eventual 30,000 acres, a capital of £600,000 was to be raised in £25 shares. Not only was that sort of money unavailable, but there were protracted difficulties in determining the rights and boundaries of the frontagers. The powers of the Company lapsed in 1858. Even so another company was floated and powers obtained by two Acts in 1879 to reclaim saltmarshes in the parishes of Freiston, Butterwick and Benington. The Crown was to be paid £4500 for the rights on the shore below high water mark, but no works were carried out and the powers of the Act expired in 1889. It was to be the last suggestion for any large-scale scheme for nearly half a century. Between the two World Wars there was another surge of ideas. In 1928 the Dynamatic Electrical Company was formed for the purpose of harnessing tides in the Wash, presumably using a barrage, and 1½ million shares of 1s were issued to the public, but it came to nought.

What also came to nought, for different reasons, was the plan for a Wash Speedway in 1929-1930. In those days of quests for speed records, it was to be England's answer to Daytona Beach. In January 1930 at Boston, the Board of Trade held a five-hour Inquiry into the application of the Automobile Racing Association (Earl Howe and Sir Alan Cobham were among its vice-presidents and it was supported by Sir Malcolm Campbell) for permission to construct a speedway for automobiles and high speed tests, a racing track, a motor boat waterway and drainage canal, and a spectator stand—all on the 'perfectly level surface of firm silt and fine mud' between high and low water mark on the Lincolnshire shore of the Wash from Gibraltar Point to Clay Hole at the mouth of the Witham. The full details of the plan are quite astonishing. The speedway itself was to be 15 miles long, 200 yards wide and surfaced with a non-skid material known as 'carpo' which contained no ingredients harmful to fish or which would contaminate the water. There would also be a twisting 12-mile TT track and a six mile motor boat speedway, an aerodrome, a racecourse stadium for dirt track, motor cycle, cycle, greyhound, pony and trotting racing, a golf course, tennis courts, bowling greens and a sand yachting course together with clubhouses, pavilions, restaurants, hostels, bungalows, camping and caravan grounds and an amusement park. Not the least incredible feature was the grandstand—a quarter of it covered—four miles long! And there would be parking for 50,000 cars.

This spectacular undertaking was to be protected from the sea by a 20-feet high bank constructed of two small and parallel lines of sandbags with silt pumped between from the main drainage canal. Some elementary experiments were conducted, in the summer (!) out on Wrangle Flats with two sandbags when it was discovered that those set end-on to the sea were disturbed less than those laid parallel to the waves. It was the opinion that when the bags rotted the sand would have become as hard as concrete! At the Inquiry it was confidently asserted that there were no known constructional difficulties 'of a technical character'. But it was not these that delayed and helped to kill the scheme, but the problem of compensating the frontagers. A total of 10,277 acres of foreshore was involved; the Association agreed terms with the Commissioners for Crown Lands and the Duchy of Lancaster, but the question of the green marsh was more difficult. At the Boston end where it was narrow, 70s an acre was reckoned sufficient compensation for loss of rights, but the frontagers with good grazing marsh at Friskney and Wrangle required £10 an acre and even then there was not complete unanimity. Moreover the shrimping and cockling fishermen of Friskney were to be deprived of their traditional living which also included the 'useful sideline' of catching sea-birds. Despite support for the speedway from Skegness and Boston, Holland County Council, drainage authorities, enthusiastic editorials in the national press, and even an attempt by farmer J. K. Measures of Skegness to build a pub called 'The Speedway' at Gibraltar Point, there was no money in the early 1930s for such a difficult and costly scheme and it was abandoned. The Campbells never came to Lincolnshire.

A rather hastily hatched scheme of a different kind was that proposed by the Dutch engineer Dr J. F. Schoenfeld in 1937. Based on the experience of damming the Zuider Zee and the creation of polders, it suggested a barrier across the Wash from Wrangle, training works for river outfalls and four 'polders' of reclaimed land. This oft-quoted Dutch solution to the problems of the Wash ignores three fundamental differences between the Zuider Zee and the Wash. The tidal range of the former was less than five feet contrasted with up to 25 feet in the Wash; the average depth of water along the Dutch barrage was 13 feet but it is some 50 feet along a line halfway into the Wash, and 85% of the bottom material of the Zuider Zee was good for cultivation whereas the saltmarshes of the Wash are but 4% of the whole.

Although Fred White first published his idea of the Wash for freshwater storage in 1949, with a barrage across the mouth of the bay, and later incorporating the Europort idea, little notice was taken of it and what debate there was sought to revive old ideas for large-scale reclamation from the sea. One aspect of the totally impractical Wash Lake suggestion which was never taken into account was the effect of increased high tide levels on the lowlying coast of central Lincolnshire. Because of the then amounts of sewage effluent pumped into the Wash, one cynic dubbed the proposal as 'the world's largest cesspool'.

Then in 1965 consulting engineers Binnie & Partners produced a report for the Ministry of Housing and Local Government—'The water resources of the Great Ouse basin'—which contained proposals for a barrage to cut off about half of the area of the Wash so that fresh water from the rivers Ouse, Nene, Welland and Witham could be pumped into two high level storage reservoirs behind the barrage. Additional benefits would be flood control, removal of the need to maintain existing sea banks, improved navigation through sea locks, availability of sites for power stations (possibly nuclear), 40,000 acres of land which would fall dry and could be extended by dumping fly ash from inland coal-fired power stations, and water recreation amenities. Although the proposal formed only a small part of the report, and followed principles for controlling fenland rivers laid down by Rennie in 1800, it caught people's imagination, and the ear of the government. The Water Resources Board saw it as one possible way of meeting future demands for water in South East England. Estimates of cost varied widely for the stages of a desk study, a feasibility study and a final scheme, phased into the next century, for four bunded (dyked) reservoirs, the last stage of which envisaged a dramatic change in the coastline between the Nene and the Welland.

The feasibility study got under way in 1972 with a budget of £2.6m. It included intensive studies by nature conservation and naturalists' organisations to try to assess the possible effect of the scheme on coastal habitats and their wildlife. It also involved the construction of a

300-yard diameter circular trial bund (water filled island) on Westmark Sand on the east side of the Nene outfall and just beyond the Lincolnshire county boundary. There it sits as a monument to a study which concluded that reservoirs could be built on the foreshore but that they are neither likely to be economic nor required this century.

Hard on the heels of the Binnie Report came the most fantastic of all concepts—that of a city of 750,000 people on the Wash. Put forward by architect Harry Teggin, it was based on the theory of cumulative gain—that is a grand scale plan for a city, a deep-water port, an airport, land reclamation and water storage which, although not viable individually, would support each other collectively and produce multiple benefits. The Great Wash City would rise from reclaimed sand and silt banks and shoals, with dock facilities in the Lynn and Boston Deeps, there would be Rennie-type agricultural 'polders' and colossal water storage reservoirs—all with road and rail links. At a stroke (or two) Lincolnshire would lose half its seaboard. A more single-minded concept, with the same effect, was that of engineer Sir Owen Williams in 1968 for a huge international airport on an island in the Wash. Simplistic in the extreme, it proposed a barrage across the mouth of the Wash which, after drainage and leaving shipping channels, would be filled in!

If such futuristic dreams were ever conceivable, the expenditure of 'your and my gold' would vastly exceed the amount of King John's treasure, even allowing for 800 years of inflation. The fact is, fortunately for the farmers, fishermen, wildfowlers, naturalists and holidaymakers who like the Lincolnshire coast as it is (and why shouldn't they?), the attitude of the English has never been to consider splashing even 1% of the gross national product into the Wash. Quite apart, that is, from the technical improbabilities; even Dutch engineers have said that a Wash barrage along the Binnie line is not feasible in the present state of our technology. Neptune is too formidable an adversary. I suspect we shall see no dramatic changes in the foreseeable future of the Lincolnshire shores of Metaris AEstuarium.

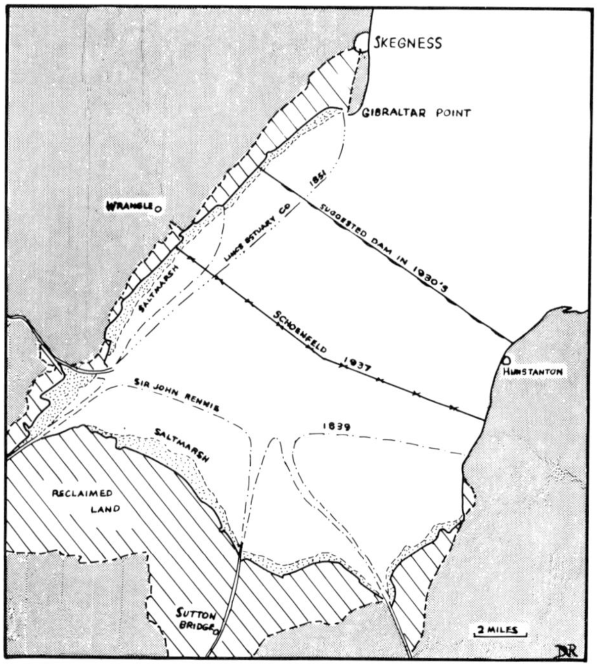

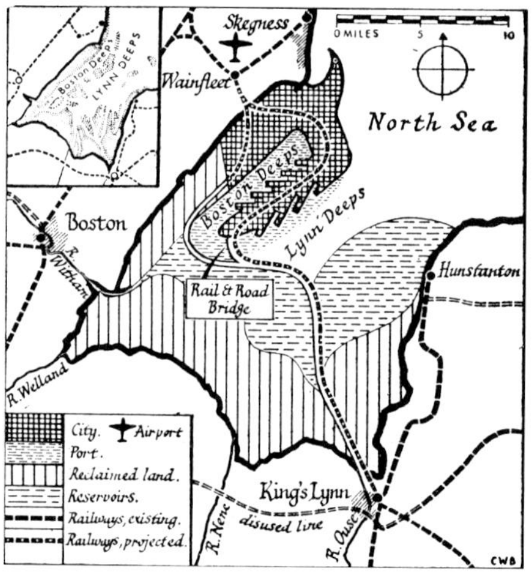

LEFT: Some failed schemes for reclamation and barrages. RIGHT: Harry Teggin's suggestion for a city in the Wash, 1966.

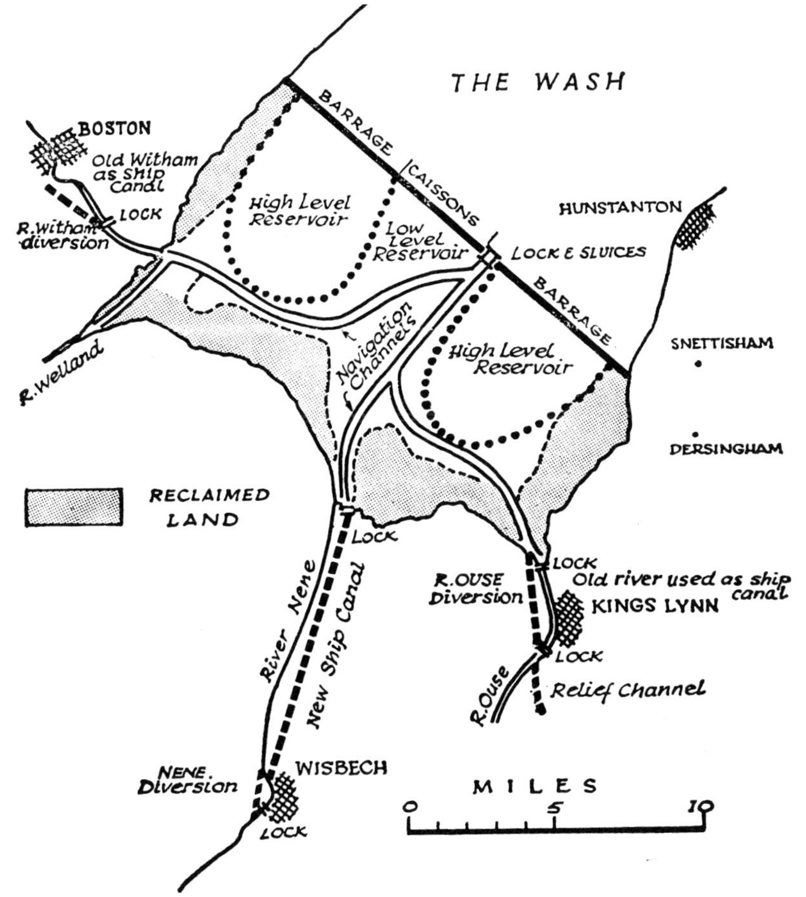

ABOVE: The Binnie proposals for a barrage and water storage scheme, 1965.
BELOW: The latest sea bank at Friskney begins to take shape, 1977-78.

Impressionist view of the Wash Speedway 1930. The four mile long grandstand highlighted in the ARA's prospectus was shown (differently) as a mile long stand, exactly opposite the 'measured mile' in the illustration accompanying the prospectus. (CES)

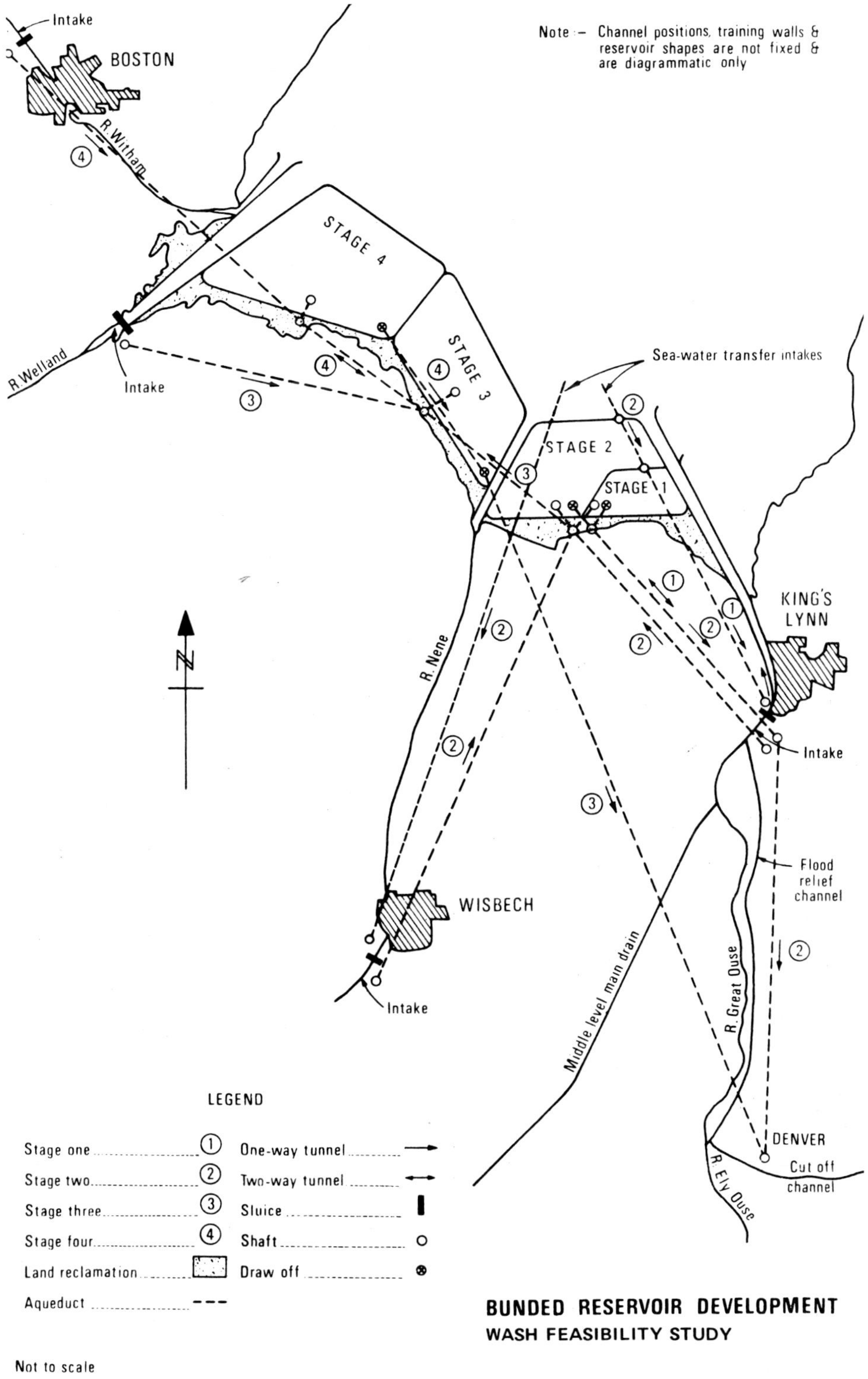

Bunded water storage reservoir proposals 1976.

ABOVE: Haile Sand fort and boom, with Tetney Haven outfall and Humberston Fitties coastline beyond. (GET) LEFT: Remains of WW1 pillbox at Gibraltar Point. RIGHT: 'Rusty pork pies' on the dunes at Sutton 1923. BELOW LEFT: Osocosy No 1 Bohemia. RIGHT: The Lodge Bungalow, Bohemia.

Conserving the Coastline

At about the time the plans for Woldsea were being unveiled, changes were afoot on the Mablethorpe sandhills—the vulnerable one of the four Ss which are the essence of the seaside (sea, sand, sandhills and salubrious air). On the Lincolnshire coast the owners of adjacent 'inlands' claimed they automatically owned the sandhills and beach above high water mark. At Mablethorpe they let some parts to the Urban District Council who re-let to amusement caterers at increasing rents as more trippers came. In 1912 the owners decided to let directly to the caterers. Alarmed at the loss of income, the Council obtained an Act of Parliament in 1914 by which it acquired ownership of the sandhills. The Council then not only challenged the amount of compensation claimed but also the title of the reputed owners, and finished up paying for arbitration, litigation and compensation totalling nearly £10,000. In one of the disputes over title before the Court of Appeal in 1918 it was adjudged that owners of adjacent lands did not have an automatic right to ownership of the sandhills, but that possessory title could be acquired as a result of exercising acts of ownership over a period of at least 12 years.

Between 1914 and 1918 the coastline was on a limited war footing. There was perhaps no more fear of invasion than there was just over a century earlier, in the Napoleonic wars, when Skegness had a beacon ready for lighting (the chain of communication was another beacon on the Wolds near Fulletby which could be seen from Lincoln). In 1914-1918 a few concrete pillboxes were built (there are the remains of one at Gibraltar Point), machine-gun posts erected (as on top of Zion Hill at Rimac) and there were beach patrols. Shipping raids were more of a danger and forts were built at the entrance to the Humber—one on Bull Sand in the middle of the estuary and the other on Haile Sand by the Tetney Haven outfall; between Spurn Point, Bull Sand fort and Haile Sand fort an anti-submarine chain was suspended, with a boom from Haile Sand fort to Humberston Fitties dunes. In 1916 an airfield was established at North Coates Fitties (on land reclaimed in 1856) where 404 Flight of 248 Squadron was based, flying Short 225 and Sopwith Baby seaplanes.

The effect of the 1918 judgement about Mablethorpe sandhills was to encourage 'inlanders' elsewhere to enclose and build on the dunes. The age of the motor car gave a new mobility to exploit the sandy seaside remote from traditional resorts and their railway stations. Bungalows, shacks, caravans and old 'bus bodies and railway carriages appeared wherever a track gave access to the sandhills and seashore. Chapel St Leonards was transformed from three farms and an inn to a flourishing seaside-oriented centre within five years, houses were built on top of the dunes at Anderby Creek and bungalows in the dunes at Soldiers Hole, and 'camps' appeared at Ingoldmells and Bohemia (the latter by 1921) together with the Trusville Holiday Camp of caravans at Trusthorpe. Bohemia, established by E. du Soleil Tupper, was an extraordinary collection of circular corrugated iron huts ('rusty pork pies') on top of the dunes near Sutton on Sea. Guaranteed 'free from convention and also restraint' with every comfort assured in 'the health-giving breeze', where 'the feeling of independence is paramount as long as one respects the few recognised rules of bohemian life', the huts were 15 ft in diameter, double-lined with stained wood floor and let furnished with electric light, water supply, Valor-Perfection oil stoves, linen, crockery and cutlery—all that was required for 'the very ideal of a Gipsy holiday life'. The rents paid for round Osocosy huts for two persons ranged from 15s a week (October-March) to £2 12s 6d in August, and the Bohemia bungalows behind the dunes (four bedrooms, bathroom and verandah) ranged up to 6gns a week in August; deck chairs were available for use on the beach at 1s a week and bathing huts for 12s 6d a week (in August), a lock-up garage was 15s a week and a store for 'motor bicycle' 5s a week. Amusements for visitors included tennis tournaments, open-air gramophone concerts and wireless listening-in. However the earth closets and poor drainage of developments like Bohemia were regarded by the County Medical Officer of Health as an acute health hazard.

The dunes were 'the longest strip of sandhills' in Great Britain and described by the Council for the Preservation of Rural England as 'sanctuaries of nature'. The dunes at Rimac were a case in point. The straightening of Saltfleet Haven in 1854 caused changes on the foreshore and a new line of foredunes was established in front of the old dunes at Rimac, completely enclosing an area which developed into freshwater marsh with reeds and sedges, water dock and marsh pennywort, bog pimpernel and marsh orchids. Reed buntings and snipe bred in the marsh, with the sedge beds as winter roosting for hen harriers. Frogs, common toads and smooth newts abounded, together with the rare natterjack toad, (in what is now its only Lincolnshire locality). Here, where grazing enclosures had been abandoned, in summer 1928 a firm of estate agents from Long Eaton in Derbyshire started to lay out Rimac-on-Sea with 188 plots for huts and bungalows—'a garden city where now is a sanctuary for birds and flowers'. Louth Naturalist Miss D. Marsden wrote in her diary that 'the caretaker walks about with a hideous plan of the estate . . . and apparently the wild fowls' home on the marshes is to be made into a lake and a row of bungalows'. Within three years 40 huts had been erected 'of varying degrees of ugliness'.

During 1926-27 two large areas of the coastline were removed from free public access with the creation of bombing and gunnery ranges by the RAF. At Donna Nook sand-dunes and strip saltings had developed on the east side of the 1863 reclamation to form The Knobs, but with sand-dunes only against the 17th century bank to the north. Beyond is a plateau-like beach of 40-50 ft thickness of silt and sand (much of it above the level of the reclaimed land) which is over a mile wide and falls relatively steeply in a series of ridges and runnels to the low water mark. The sand flats and bars provide a hauling-out area for grey and common seals (both pup here) and support one of the few breeding colonies of little tern on the Lincolnshire coast. About 2,500 acres of these natural features above the high water mark of ordinary tides (as surveyed in 1905 and now an arbitrary line without real meaning, although still slavishly copied by atlas and other maps)were taken over for three bombing targets (one illuminated for night exercises) and ten gunnery targets. The development was not exactly welcomed by local residents, particularly when smoke bombs missed their target by 1½ miles. The other range was (and still is) in the Wash on the saltmarshes and mudflats of Holbeach St Matthew, aircraft flying from Sutton Bridge for bombing and gunnery training.

More squatters enclosing the sandhills and erecting 'eyesores and abominations' brought to a head the question of traditional public access through the dunes to the beach.Triggered off by a specific case of a hotel car park on unenclosed land at Sutton on Sea, Lindsey County Council began an investigation of the whole question of ownership rights of the sandhills in 1927. The result was the promotion of a Parliamentary Bill in 1930 to grant powers to preserve all the sandhills and beaches from Donna Nook to Gibraltar Point as open spaces and to regulate future development. Although petitions against it were eventually withdrawn, the Bill was regarded as 'so novel in character' that it was treated as an Opposed Bill. The plan was first to gain control of development through licences. Later, relatively small parts of the sandhills might be acquired, with compensation and by compulsion if necessary, but only for the first five years after the Act. The Bill was seen by one local newspaper as 'the nearest approach to a scheme for land nationalisation since the Norman Conquest', but most people saw the advantages of regulating development before the sandhills disappeared in an unsightly sprawl of shacks. A compromise was reached with petitioners from Skegness, £4,750 was paid in compensation for disturbance at Bohemia, and the far-sighted Bill was given the Royal Assent in July 1932. By 1937, 17 controlled areas had been designated (compensation £5,107); 48 new huts were on the controlled hills at Mablethorpe and Sutton, and 30 at Anderby Creek, Chapel St Leonards and Ingoldmells. (The sea defences round the sandy promontories covering drain outfalls at Chapel and Ingoldmells then consisted of oak posts driven into a clay toe to the dunes and between them a basketwork of ash poles—a simple but effective method of absorbing wave energy.) The effect of the Act had been to stop undesirable development, while at the same time about 500 acres of land were purchased (from about 130 owners) for over £35,000. The camp sites at Huttoft and Ingoldmells were acquired and caravans evicted; today there is a splendid car terrace at Huttoft Bank and a model caravan site run by the County Council at Ingoldmells Point.

Land at Gibraltar Point in the parish of Croft had been sold off in 1923 from the extensive Tyrwhit-Drake estates to Mr G. Hyams and Mr J. Giles of Skegness who proposed to develop a 'garden city'. In 1930 just over 200 acres of sandhills and saltings were bought from Hyams and Giles by MacLean Estates to develop the Tennyson Glen estate of 800 houses, with promenade, greens, courts and shelters, and building commenced. Only three houses were built (and are still there on Aylmer Avenue) before the company went into liquidation in 1934 and Lindsey County Council purchased the land for £12,000. The National Trust agreed that the area was 'of considerable beauty and great interest from the botanical and ornithological points of view' and recommended the financial support of the Pilgrim Trust (which would have been appropriate in view of the Lincolnshire connections with the Pilgrim Fathers—sailing from Skirbeck near Boston in 1607 and Killingholme Haven in 1608). In the event the Ministry of Health sanctioned borrowing of the purchase money (because of extraction of sand from the western dunes). At Saltfleetby-Theddlethorpe the RAF had established a third bombing and firing range between 1932 and 1936 by purchase of dunes from 65 owners, including 41 at Rimac and Mablethorpe UDC, and foreshore above the 1905 HWMOT. In 1938 another bombing range was set up on the marshes and mudflats at Wainfleet on the site of an artillery range of the 1890s.

During the second World War the whole of the coastline was taken over by the armed services. Gibraltar Point was used for military training, with a rifle range using moving targets towed along a narrow gauge railway (the track bed and engine shed are still there in the eastern dunes) and a mortar range on the beach. Fragments of shell casing still come up among the beach stones. Butlins Holiday Camp became HMS Royal Arthur, and RAF North Coates was taken over by Coastal Command, attacking enemy shipping with Bristol Beauforts and Beaufighters and Lockheed Hudsons. Coastal fortifications included 6-inch gun batteries at Grimsby, Mablethorpe, Jackson's Corner (Winthorpe), Gibraltar Point and Freiston Shore (where the concrete works are still in a reasonable state of preservation). A considerable number of rectangular and hexagonal concrete machine-gun pillboxes—some camouflaged as sheep pens or pig styes—were built into the dunes or on top of the sea wall at regular intervals and strategic points. Heavy ordnance positions north of Theddlethorpe were of brick with concrete and steel roof. Lines of 4-ft cube concrete blocks were laid down for anti-tank defence (a number survive at Gibraltar Point, Saltfleetby and Donna Nook), and on the open beaches a framework of iron poles and rolls of barbed wire, not all cleared away after the war as they tended to trap wind-blown sand and in some cases became incorporated into the dunes.

Shortly after the war, in 1946, an ambitious scheme was prepared by London architect G. A.

Jellicoe to develop a promenade from North End pullover, Mablethorpe to Sandilands with dunes providing a natural background. Sections of the sandhills would be developed for a park with chalets and tree-lined walks, a grand pavilion with theatre and restaurants, a bathing pool, and canal with water 'bus service behind the dunes. Only the North Dunes Gardens and walks with a few chalets were built.

Unfortunately the dunes at Humberston Fitties (which had built up in front of Anthony's Bank between 1795 and 1830) were not included in the pioneer Sandhills Act. Between the wars a number of huts were erected on the dunes and the Humberston Fitties Campers'Association was formed in 1925. By 1947 there were about 60 huts and bungalows, and 240 more were erected in the next nine years, and the dunes were already in a degraded state from the effects of visitor trampling. Where the dunes had extended southwards, a small area of marsh behind was reclaimed in 1941 and later developed as a caravan site (now over 1,300 vans) further increasing the pressure on the dunes. After the 1953 Storm Flood, kidding, fencing and marram grass planting was undertaken, along with some conifer planting. These measures were inadequate to combat increased wave action as the outfall of Tetney Haven changed its course across the foreshore and gave deeper water parallel to the dunes. (The World War I boom to Haile Sand fort, where the outfall used to reach low water, was removed in 1965-66.) Following half-hearted attempts to deal with erosion of the dunes, an£18,000 gabions (mesh basket) wall was built. However, the next two winters' tides revealed a serious defect as sand was scoured from the beach and the filling of slag slipped out of the bottom of the steel mesh basket on to the beach; the basket still needs topping up. The construction of groynes has helped to alleviate the problem, but the whole story of shortsighted and haphazard development of sand-dunes and saltmarshes is a sad contrast with other areas of the Lincolnshire coast.

Elsewhere the Sandhills Act has enabled the County Council to control the demand for caravan and camping facilities which Lindsey County Council had correctly forecast 20 years before. The number of caravans between Mablethorpe and Skegness rose from 4,200 in 1960 to 11,000 in 1959. The bulk are static caravans; the wheels have little practical purpose and it was significant that one site at Anderby Creek was named the Poplar Zareba (where Zareba is Arabic for 'where my caravan has rested'). Today there are over 19,000 zarebas, more than a third at Ingoldmells, often dubbed 'caravan city'. There are also 2,500 chalets including Humberston, with the bulk at Mablethorpe and Sutton, on sites rejoicing in such unlikely names as Miami Beach. The still-accreting foreshore at Skegness enabled development of amenities even further out from the original parades with the construction of the Bracings in the 1950s. However the present vulnerability of the dunes towards the north end of Skegness front raises the question of how long that accretion will continue, as the pattern of the foreshore inevitably changes.

It is the changing pattern of the past which has created Gibraltar Point. Along much of the Lincolnshire seaside the veneer of sand over the post-glacial clays is relatively thin. The bulk is derived from offshore sources (with only a small amount from Holderness) by tidal current movement, which produces a pattern of flood and ebb channels with ridges between, such as the Skegness Middle, Inner and Outer Dog's Head and Lynn Knock off Gibraltar Point. Once emplaced on the foreshore, the sand is subject to wave action. With variations of gradient according to the grain size of the sand, wave action produces four basic ridges on the beach—at the levels of high and low water spring and neap tipes—with runnels between (which are sometimes deep enough to reveal the underlying clays, peat and submerged forest). As the waves approach the beach at an oblique angle from the north east, wash and backwash of the waves gradually impells the beach material in a southerly direction—longshore drift. Within the Humber estuary the drift from Donna Nook is into the estuary. It is not only the individual grains of sand which move, but in fact the whole system of beach ridges is on the move. One of the effects of longshore drift is to divert the outfalls of rivers; this was the reason why Tetney Haven changed its outfall towards the north-west.

Similarly the growth of the sandy spit and dunes southwards from Skegness in the last 600 years or so has increasingly diverted the outfall of the Steeping River or Wainfleet Haven. The dunes which now form the Seacroft Golf Course and western dunes at Gibraltar Point extended as a spit, to approximately their present limit by the third quarter of the 18th century, but it took some time before the terminal end was consolidated. This was not aided by the digging of gravel which allowed the sea to break up the spit into a series of islands by 1824. To the east was open foreshore. Visitors to Gibraltar Point in the 1860s described it as 'a sort of Lands End Place'. Beyond the Old Ship Inn and the newly erected Coastguard Station (1859) 'what a strange desolate-looking scene it is: a broad flat patch rough with marum and sea-thorn, stretching out upon a scene yet more desolate, broad leagues of rough, wet, ripply sand'. Yet, advancing slowly from the north, was a complex of beach ridges and dunes within the shelter of which finer silt settled and saltmarsh developed. The line of parish boundary (between Croft and Wainfleet St Mary) across this new-growing foreshore was marked by iron posts bearing the initials of the respective lords of the manor—T. T. D. (Thomas Tyrwhit Drake) and Ld B. (Lord Bayning).

One of the first plants to colonise the mud of a new saltmarsh is glasswort or marsh samphire ('samfer'). According to Gerard's Herball (1636) 'glasswort hath many grosse thicke and rounded stalks . . . full of fat and thicke sprigs, set with many knots or joints, without any leaves at all.' It is a Lincolnshire delicacy. It should be picked when dark green and shiny. The traditional gathering day was the old August Monday; young shoots pinched off above the roots in June make a crisp and tangy salad vegetable. In July and August the whole plant should be pulled, roots and all. To serve as an asparagus-like starter, wash well and boil upside down in a saucepan of water. Drain and serve in a bowl with molten butter. The art of eating it is to tilt the head back, pick up by the stem at the root end, pop in the mouth and draw off from the ribs the fleshy part with the teeth. For a really traditional Lincolnshire dish, samphire needs to be pickled—and eaten with stuffed chine.

By the time the sheltering eastern dunes line had reached its present limit before the end of the 19th century, much of the saltmarsh had matured from a spread of samphire and seablite through sea meadow grass and scurvy grass to sea aster and sea lavender with the shrubby sea purslane along the banks of creeks, and the aromatic sea wormwood where the tide reached less frequently. In 1922 a storm beach was thrown up across the southern edge of the marsh, with one main creek for the sea to flood in and out. The summer carpet of sea lavender, with its slightly varying shades of purple, is the most extensive on the Lincolnshire coast. Bulldog Bank was raised between the two line of dunes at the north end of the marsh in the 19th century, and the freshwater marsh thus created was grazed.

On the outer side of the eastern dunes a series of crescent-shaped beach ridges developed and became stabilised as dunes, with strip saltings between. This is still a mobile and developing part of the shoreline; dune ridges can be combed down and new beach ridges built up. In the last 30-40 years a new spit has extended south beyond the end of the eastern dunes and a new marsh has grown up within its shelter. New dunes, colonised by sea couch grass and then marram, may be host for a time to handsome rarities like the sea holly, but are likely to be invaded before long by sea buckthorn—its bright orange berries in autumn attracting flocks of migrating fieldfare. The thick scrub of sea buckthorn on the older dunes is the most extensive on the east coast of England.

Lindsey County Council's Sandhills Act and the failure of MacLean Estates had saved Gibraltar Point, but little could be done until after the war, although in 1939 the Lincolnshire Naturalists' Union had suggested that a bird observatory be established at Gibraltar Point. In 1948 the Lincolnshire Naturalists' Trust (now the Lincolnshire and South Humberside Trust for Nature Conservation) was formed; through the drive and initiative of its honorary secretary Ted Smith (then University of Nottingham Resident Tutor for South Lindsey) the County Council quickly agreed to the establishment of a nature reserve. Later an area of dunes and foreshore owned by the then Skegness UDC was added. In 1952 the County Council designated their part of the reserve as the first Local Nature Reserve in England and Wales under the National Parks and Access to the Countryside Act of 1949.

From its inception as a nature reserve, the value of Gibraltar Point for education and research was stressed. (Back in 1907 it had been suggested that 'the physical geography of the Lincolnshire sandhills might be usefully taught in the schools of Lincolnshire and the Midlands'.) A large Heligoland bird trap was built by the time the bird observatory opened in an old army building in the eastern dunes, in April 1949; 1,564 birds, of 61 species, were recorded in the first eight months. Physiographic and botanical research was undertaken by the University of Nottingham, and still continues, having added considerably to our understanding of coastal processes. In 1958 Lindsey CC acquired the old coastguard station (two extensions had been built in the 1930s when it was used as a holiday home) and leased it to the Trust as a field research station. Laboratories and dormitory accommodation were added in 1964 to enable residential field teaching for Lincolnshire and other schools, for adult courses by the University of Nottingham's Department of Adult Education and for use by amateur naturalists. A small information centre was also built on the main car park. By the 1970s the number of day visitors to the reserve had built up to 200,000 a year and in 1974 the County Council, with grant-aid from the Countryside Commission and the Carnegie UK Trust, built the Visitor Centre with an exhibition and sales area; a freshwater mere, complete with viewing hide, was also created.

With its dynamic complex of sand-dunes, saltmarshes, sandy beaches and mud flats, Gibraltar Point is recognised for its international scientific importance. The Trust has done pioneer work in the protection of the shore-nesting ringed plover and rare little tern. As the autumn high tides cover the feeding grounds on the mud flats, flocks of oystercatcher, godwit, knot and dunlin rise in huge clouds to take refuge on the spit and sand-dunes. The successful protection of this part of the Lincolnshire seaside and the conservation of its fragile and diverse habitats, while at the same time ensuring that the large number of visitors do not damage what they come to enjoy—and enabling a programme of education at different levels—is a fine example of the successful collaboration between a local authority and a voluntary body. It can still be said of Gibraltar Point, as it was in 1897, that 'it is ideal for the lover of nature in its wildest beauty.'

There are also other nature reserves on the Lincolnshire seaside. In 1956 the Trust was able to establish a nature reserve where Rimac-on-Sea would have been; when the bombing range was moved to Donna Nook the Nature Conservancy Council purchased most of the land from the Ministry of Defence and it was declared the Saltfleetby-Theddlethorpe Dunes National Nature Reserve in 1968. Some 2,400 acres of dunes, strip salting and foreshore at Donna Nook, extending along 6½ miles of coastline from Somercotes Haven to Skidbrooke, is now managed as a nature reserve by agreement between the Ministry of Defence and the Trust. On this reserve 152 species of plants have been recorded, and 47 species of bird breed regularly. Grainthorpe Fitties to the north is a site of Special Scientific Interest, and at Tetney/North Coates the Royal Society for the Protection of Birds leases a 2,500-acre reserve of marsh and dune foreshore to protect a large breeding colony of little tern.

The concepts of coastal conservation stemming from the Sandhills Act, outside the main holiday concentrations of Mablethorpe-Sutton and Ingoldmells-Skegness, have been to provide suitable opportunities for informal recreation while at the same time conserving the natural features of the coastline. The general vulnerability of those natural features emphasises the need to understand the conflicts and dangers which can be caused by recreational over-use of the seaside.

The saltmarshes of the Wash are a different problem. They are part of the Grade 1* Site of Special Scientific Interest of international significance which covers the whole Wash. There is a 670-acre nature reserve on Frampton Marsh, some of the oldest saltmarsh in the Wash, leased by the Trust from the Anglian Water Authority, and there is conservation involvement with the marshes on the Holbeach Firing Range. The boundary of the Wainfleet Firing Range has been adjusted to accommodate the latest saltmarsh reclamation for agriculture, and there is pressure to reclaim more of the existing marsh on the thesis that the new marsh will always grow rapidly outside a new reclamation bank, as it has in the past. That it will continue to happen in the future is much less certain.

Although coarse sandy material comes ashore north of Gibraltar Point, and forms much of the bed of the Wash, very little new fine mud suitable for saltmarsh formation enters the Wash now either from the sea or down the rivers (which have long floodplain courses and are sluiced at the seaward exits). Thus each new reclamation involves the permanent withdrawalof partof the near finite circulating supply of mud. Moreover the position of low water mark has changed very little by accumulation of coarse sediments in the last 150 years, during which sea level has risen by over a foot, whereas the high water mark has been advanced by reclamations much more in the same period.Consequently the gradient of the foreshore has steepened and the intertidal zones of saltmarsh, mud and sand flats have become narrower. Certainly the 20th century reclamations have been smaller than those of 300 years before. And because of the rising sea level it means that each successive reclamation is slightly higher than the previous one (with obvious long-term implications for gravity drainage).

If the intertidal zones of the Wash coast were to be further truncated by continued saltmarsh reclamations, there would be serious consequences for wildlife. The saltmarsh and mudflats provide breeding and feeding areas for very large numbers of birds. Only three other areas of Europe support more wintering waders, with something like a quarter of a million visiting the Wash in a year, and 40,000 twite (a species of linnet), feed on Wash saltmarshes in the winter. Of the 20 species nesting regularly in the saltings, redshank (over 1,250 pairs), skylark, meadow pipit and read bunting form a substantial proportion. It is significant that the higher and older parts of the saltings have the greatest variety of species and the greatest density of birds. In 1980 an application to reclaim saltmarsh at Gedney Drove End was the subject of a Public Inquiry; the outcome was refusal of planning permission by the Secretary of State for the Environment who concluded that ecological and scientific considerations should prevail over immediate agricultural needs. Nevertheless one suspects that the future of the saltmarshes of the Lincolnshire seaside is not yet secure.

Of paramount significance for the whole Lincolnshire seaside is protection from the sea. The maintenance of the sea defences—reclamation banks, sandhills and sea walls—is now the responsibility of the Anglian Water Authority. For the future protection of agriculture and the holiday industry it will be necessary to heighten or strengthen all Lincolnshire's sea defences, and more land drainage will have to be pumped out to sea. It is not a problem that will go away. The Lincolnshire seaside is still sinking.

OPPOSITE ABOVE: 'Bungalows' and shacks in the dunes at Rimac 1931, (LCC) and BELOW: The same area today—now the Saltfleetby-Theddlethorpe Dunes National Nature Reserve. Open area in foreground is where the 'bungalows' were sited. CENTRE: Seals hauled out at Donna Nook. LEFT: Bombing target on the former Theddlethorpe range. RIGHT: Shrimping at Chapel St Leonards, 1914. ABOVE: The car park at Sutton which triggered off Lindsey County Council investigation of sandhill ownership rights. (LCC) BELOW: Huts in narrow and vulnerable dunes at Sandilands 1930. (LCC)

ABOVE: Squatters' bungalows built on top of the sandhills at Soldiers Hole, Chapel St Leonards 1930. (LCC) CENTRE: Hollands Gap, Chapel St Leonards, with tents, 'bus bodies and railway carriages, 1930. BELOW: The basketwork sea defences of Chapel Point 1920—note also the WW1 pillbox.

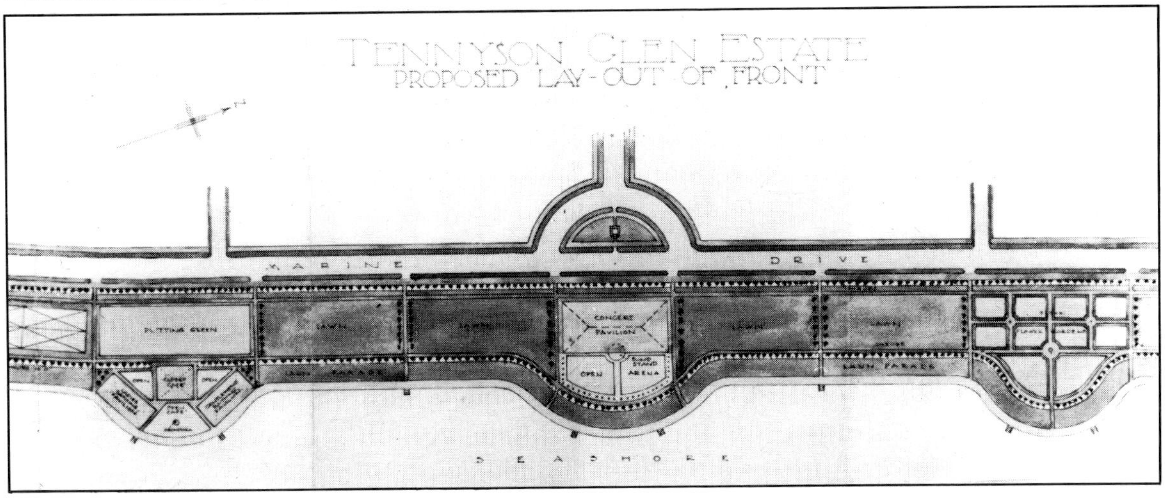

LEFT: A modern eyesore at Ingoldmells, 1960s. RIGHT: New bungalows being built by contractor E. Tinkler—now Aylmer Avenue, Gibraltar Point. (LCC) CENTRE: Tennyson Glen Estate: proposed layout of seafront at Gibraltar Point. BELOW: Degraded dunes in front of the chalet camp at Humberston Fitties; the position of the crazily-leaning WW2 pillbox shows how much of the sandhills has been eroded by the sea, (1962).

LEFT: The wire basket gabions wall with rubble falling out of the bottom hardly makes an attractive holiday beach, 1978. RIGHT: Fishing boats in the silting Wainfleet Haven today. BELOW: Golden Sands caravan site, Mablethorpe.

ABOVE: Gibraltar Point from the air, 1960. (A) BELOW: Noah's Ark Hotel, Gibraltar Point, selling ginger ale, lemonade and nectar, 1897.

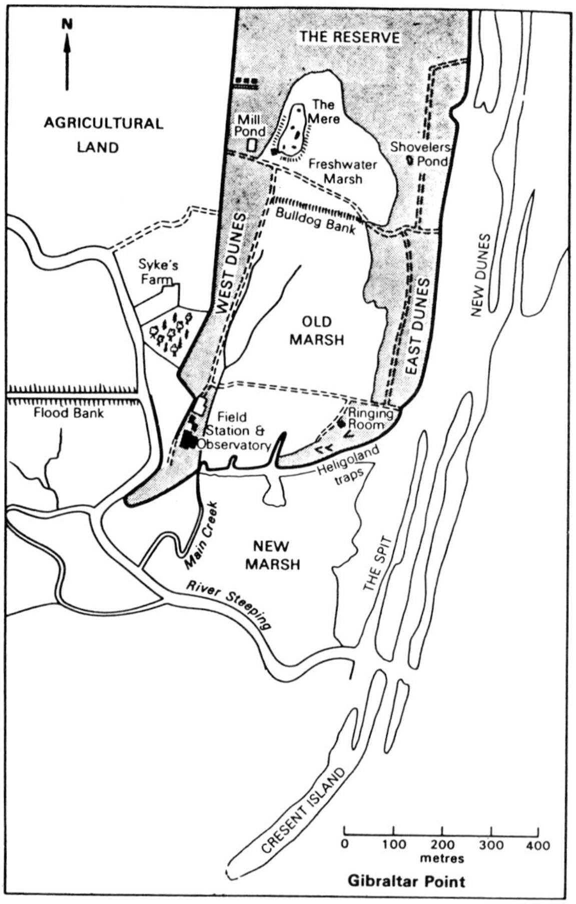
N
THE RESERVE
AGRICULTURAL LAND
Mill Pond
The Mere
Shovelers Pond
Freshwater Marsh
Bulldog Bank
Syke's Farm
WEST DUNES
OLD MARSH
EAST DUNES
NEW DUNES
Flood Bank
Field Station & Observatory
Ringing Room
Heligoland traps
Main Creek
NEW MARSH
THE SPIT
River Steeping
CRESENT ISLAND
0 100 200 300 400
metres
Gibraltar Point

Gibraltar Point. OPPOSITE LEFT: Nature reserve. RIGHT: Army wheel tracks, 1940s, in the saltmarsh; extra saline conditions in the ruts inhibit the growth of sea lavender, and BELOW: the Coastguard Station, built in 1859, seen here c1925. CENTRE: The Field Station today, built round the old coastguard station. (RBW) BELOW: The Visitor Centre. ABOVE: Visit of Prince Charles to the Nature Reserve 14 July 1971 with left to right: the author (Hon Secretary of the County Trust for Nature Conservation), Paul Berkeley, Ted Smith (Chairman of the Trust), Walter Lane (then Clerk of Lindsey County Council, now President of the Trust). BELOW: The sleeper track across the eastern dunes at Gibraltar Point which stops erosion by trampling.

ABOVE: Extensive samphire beds on the muddy shore at Skidbrooke. LEFT: The almost too-well camouflaged ring plover chick and three eggs. RIGHT: The natterjack toad. (RMR)

ABOVE: University of Nottingham adult education courses at Gibraltar Point include 'Birds in autumn'. (GET) BELOW: The view across the 12-mile wide Wash to Hunstanton on the Norfolk coast.

Bibliography

Pebbles on the Beach

Cameron K. (ed) *Place-Name Evidence of the Anglo-Saxon Invasion and Scandinavian Settlements* (Nottingham 1977)

Jukes-Brown A.J. *The Geology of East Lincolnshire* (London 1887)

Phillips C.W. The Roman Ferry across the Wash *Antiquity* 4, 1932, 342-

Phillips C.W. (ed) *The Fenland in Roman Times* (London 1970)

Read C. *The Geology of Holderness and the adjoining parts of Yorkshire and Lincolnshire* (London 1885)

Robinson A.H.W. The Inshore Waters, Sediment Supply and Coastal Changes of part of Lincolnshire *East Midl. Geogr.* 3, 6, 22, 1964, 307-321

Robinson A.H.W. The Submerged Glacial Landscape of the Lincolnshire Coast *Trans. & Papers Inst. Brit. Geogrs.* 44, 1968, 119-132

Robinson, D.N. Some physical aspects of the East Coast of Lincolnshire (unpub. dissert. University of Nottingham, 1952).

Robinson D.N. Pleistocene Deposits in Lincolnshire—excursion report *Mercian Geologist* 5, 4, 1975, 337-339

Robinson D.N. Gibraltar Point, Lincolnshire & the Northern Fen Edge—excursion report *Mercian Geologist* 6, 2, 1977, 133-135

Serra J.C. de On a Submarine Forest on the East Coast of England *Phil. Trans.* 89, 1799, 145-156

Simmons B.B. Ancient Coastlines around the Wash *S. Lincs Archaeol.* 1, 1977, 6-9

Straw A. Pleistocene Events in Lincolnshire: a survey and revised nomenclature *Trans. Lincs. Nats. Union* 17, 2, 1969, 85-98

Swinnerton H.H. The Post Glacial Deposits of the Lincolnshire Coast *Quart. Journ. Geol. Soc.* 87, 2, 1931, 360-375

Swinnerton H.H. The Physical History of East Lincolnshire *Trans. Lincs. Nats. Union* 1936, 91-100

Swinnerton H.H. & Kent P.E. *The Geology of Lincolnshire* 2nd ed. (Lincoln 1976)

The Lost Towns and **The New Lands**

Boyton P. Jean Ingelow: High Tide on the Coast of Lincolnshire *Lincs. Poacher* 1, 3, 1953, 11-15

Buchner G.E.E. *A Reclamation near Wainfleet, Lincolnshire* Inst. Civ. Eng. Maritime & Waterways Paper No. 12 (London 1949)

Hallam H.E. *The New Lands of Elloe* Occ. Paper No. 6 (Univ. Leicester 1954)

Ingliss C.C. & Kestner F.J.T. Changes in the Wash as affected by training walls and reclamation works *Proc. Inst. Civ. Eng.* 11, 1958, 435-466

Kestner F.J.T. The Old Coastline of the Wash *Geog. Journ.* 128, 4, 1962, 447-478

Owen A.E.B. Coastal Erosion in Lincolnshire *Lincs. Historian* 9, 1952, 330-341

Owen A.E.B. The Roman Bank *Lincs. Poacher* 1, 3, 1953, 25-27

Owen A.E.B. The Early History of the Saltfleet Haven *Lincs. Architect. & Archaeol. Soc. Reports & Papers* 5, 2, 1954, 1-14

Owen A.E.B. Wilgrip Haven and Theddlethorpe *Lincs. Historian* 2, 3, 1955/56, 37-41

Owen A.E.B. The Levy Book of the Sea: the organisation of the Lindsey Sea Defences in 1500 *Lincs. Architect & Archaeol. Soc. Reports & Papers* 7, 1957, 35-48

Owen A.E.B. The Upkeep of the Lindsey Sea Defences 1550-1650 *Lincs. Historian* 2, 10, 1963, 23-30

Owen A.E.B. Hafdic: a Lindsey Name and its implications *Journ. English Place-Name Soc.* 7, 1974/75, 45-56

Robinson D.N. The North-East Coast of Lincolnshire: A Study in Coastal Evolution (unpub. MSc thesis Univ. Nottingham 1956)

Robinson D.N. Coastal Evolution in North-East Lincolnshire *East Midl. Geogr.* 5, 1/2, 1970, 62-70

Smith G.I. *L.U.S. Report 69 LIncolnshire* (Parts of Holland) (London 1937)

Smith P. *The Story of Claribel* (Lincoln 1965)

Thirsk J. *English Peasant Farming: The Agrarian History of Lincolnshire from Tudor to Recent Times* (London 1957)

Wheeler W.H. *History of the Fens of South Lincolnshire* (Boston 1868 & 1896)

Hydraulics Research Station *The Wash Storage: A Historical Review* Report DE26 (Wallingford 1975)

Salt in Season

Baker F.T. The Iron Age Salt Industry in Lincolnshire 26-34

Hallam H.E. Saltmaking in the Lincolnshire Fenland During the Middle Ages 85-112

Hallam S.J. The Romano-British Salt Industry in S. Lincolnshire 35-75

Rudkin E.H. & Owen D.M. The Medieval Salt Industry in the Lincolnshire Marshland 76-84

. . . *Lincs. Architect. & Archaeol. Soc. Reports & Papers* 8, 1959-60

de Brisay K.W. & Evans K.A. *Salt: The Study of an Ancient Industry* (Colchester 1975)

Alvey R.C. Post-glacial fauna and flora from intertidal exposures in the Ingoldmells area *Mercian Geologist* 3, 2, 1969, 137-142

Dover D. *Salt* (Oxford 1975)

Healey R.H. Medieval Salt Making *South Lincs. Archaeol.* 1, 1977, 4-5

Longley Rev T. Paper on Lincs. Marshes Salt Industry *Louth Adverter* 7 Apl 1900

Maudson Grant S. Ancient Pottery Kilns *Lincs. Notes & Queries* viii, 62, 1904, 33-38

May J. *Prehistoric Lincolnshire* (Lincoln 1976)

Robinson D.N. Beach Excavations at Ingoldmells Point *Survey* 5, 1, 1954, 42-45

Simmons B.B. *The Lincolnshire Car Dyke* (Swineshead 1975)

Sturman C.J. The Salt Industry of the Lincolnshire Coastal Marshlands (unpub. paper, Exeter 1976)

Sturman C.J. & Robinson D.N. Settlement & Landscape History *in* Soils in Lincolnshire TF39 Soil Survey Record (in press)

Swinnerton H.H. The Prehistoric Pottery Sites of the Lincolnshire Coast *Antiquaries Journ.* xii, 3, 1932, 239-253

Walshaw G.R. An Ancient Lincolnshire Map *Lincs. Mag.* 2, 7, 1935, 196-206

Whitwell J.B. *Roman Lincolnshire* (Lincoln 1970)

The Smugglers
Burton C.D. Farewell to the Smugglers *Lincs. Life* 8, 1, 1968, 33
Daniels S. Wool Smuggling in Lincolnshire *Lincs Life* 14, 12, 1975, 28-29
Perkins J.A. The Lincolnshire Contraband Tobacco Trade after the Napoleonic Wars *Journ. Transport History* 4, ii, 1977, 86-101
Perkins J.A. *Sheep Farming in the 18th & 19th Century in Lincolnshire* (Lincoln 1977)
Swaby J.E. *The Marshmen* (London 1962)
Walter J.C. *Mablethorpe: Its History and Associations with appendix of Smuggling Anecdotes* (Louth 1907)
Webb W. *Coastguard! An Official History of H.M. Coastguard* (London 1976)
Wood G.B. *Smugglers' Britain* (London 1966)

Neptune's Ale
Dobson E. *A Guide and Directory to old Cleethorpes* (Cleethorpes 1850)
Hern A. *The Seaside Holiday* (London 1967)
Howell S. *The Seaside* (London 1974)
Preston H. Report of field Meeting, Freiston shore *Trans. Lincs. Nats. Union* 1921, 160-161
Rawnsley H.D. *Memories of the Tennysons* (Glasgow 1900)
Sturman C.J. The development of sea bathing in Lincolnshire in the late 18th and early 19th centuries (unpub. MS)
Tower Rev E. Richard Fowke's Journey to Freeston Shore *Assoc. Architect. Socs. Repts.* 1867/68, 364-369
White W. *Eastern England from the Thames to the Humber* 2 Vols (London 1865)

Day of the Tripper and **Beside the Seaside**
Aspinall P.J. Speculative Builders and the Development of Cleethorpes, 1853-1900 *Lincs. Hist. & Archaeol.* 11, 1976, 43-52
Baker F. *The Story of Cleethorpes* (Cleethorpes 1953)
Beastall T.W. *A North Country Estate: The Lumleys and Saundersons as Landowners 1600-1900* (Chichester 1975)
Bolland W.T. Places of Interest on the Great Central—Cleethorpes *Grt. Cent. Rlwy. Journ.* June 1906, 280-283
Dow G. *The Alford & Sutton Tramway* (Lingfield 1947)
Dutton G.H.G. *Ancient & Modern Skegness & District* (Skegness 1922)
Faulds I. The Vine Hotel, Skegness *Lincs. Life* 13, 8, 1973, 58-59
Gurnham R. The Creation of Skegness as a Resort Town by the Earl of Scarbrough *Lincs. Hist. & Archaeol.* 7, 1972, 63-76
Hackett E. A Jungle in Skegness *Lincs. Life* 20, 5, 1980, 32-33
Hallgarth S. The White House, Ingoldmells *Lincs. Life* 19, 2, 1979, 30
Kime W. Forty Years Ago at Skegness *Lincs. Life* 8, 7, 1968, 30
Kime W. *Skeggy! The Story of an East Coast Town* (Skegness 1969)
Kime W. Skegness Centenary 1877-1977 *Lincs. Life* 17, 5, 1977, 18-24
Kime W. *Bygone Skegness* (Grantham 1980)
Lamming D. Trips to Skeggy—Twenties Style *Lincs. Life* 17, 5, 1977, 28-29
Manning P. *Looking at Mablethorpe* (Louth 1952)
Martineau H.D. Skegness is so Bracing *Lincs. Life* 9, 6, 1969, 12-20
Martineau H.D. Colourful Cleethorpes *Lincs. Life* 13, 6, 1973, 25-35
Murray's Handbook for Lincolnshire (London 1900)
North R. *The Butlin Story* (London 1962)
Patchett T.S. A History of Mablethorpe & District (unpub. MS)
Pearson R.E. The Lincolnshire Coast Holiday Region (unpub. MA thesis Univ. Nottingham 1965)
Pearson R.E. Railways in Relation to Resort Development in East Lincolnshire *East Midl. Geogr.* 4, 5, 1968, 281-294
Pimperton R.S. Ingoldmells *Lincs. Life* 19, 2, 1979, 24-29
Robinson D.N. Beside the Seaside *This England* 1, 2, 1968, 14-19
Shannon I. Cleethorpes *Lincs. Life* 5, 1, 1965, 31-40
Ward Lock Pictorial & Descriptive Guide to Skegness, Sutton, Mablethorpe & Woodhall Spa (London 1907)
White P.R. Rails to Skegness *Lincs. Transport Rev.* 7, 7, 1969, 105-109; 8, 8, 1970, 113-117; 9, 1, 1971, 4-10
South East Lindsey Teachers' Centre *Discovering Skegness* (Skegness 1980)
—*Seaside Watering Places* 1899-1900 (London)

Piers and Pierrots
Adamson S.H. *Seaside Piers* (London 1977)
Bratley M.A. Clements Arcadia *Lincs. Life* 8, 6, 1968, 38-39
Hannam S.D.J. & Robinson D.N. The End of the Pier? *Lincs. Life* 18, 11, 1979, 18-24
Robinson D.N. The Clements Story *Lincs. Life* 15, 7, 1975, 32-36

Launch the Lifeboat
Davison A.J. *Britain's Life-boats* (London 1923)
Gibney J. H.M. Coastguard *Lincolnshire Life* 11, 10, 1971, 50-54
Jones G.P. *The 'Anzio I' Disaster* (1966)
Major F.S.W. *A Century and a Half of Skegness Lifeboats* (Skegness 1977)
Storey A. *Hull Trinity House History of Pilotage and Navigational Aids of the River Humber 1512-1908* (Hull 1971)

The Great Storm Surge of 1953
Barnes F.A. & King C.A.M.
The Lincolnshire Coastline and the 1953 Storm Flood *Geography* xxxviii, 3, 1953, 141-160
Barnes F.A. & King C.A.M.
Notes on the Causes of the Recent Tidal Inundations round the North Sea *Survey* 3, 2, 1953, 29-36
Critchley A.V.G. East Coast Flood *Lincs. Life* County Annual 1965, 76-78
Robinson A.H.W. The Storm Surge of 31 January-1 February 1953 *Geography* xxxviii, 3, 1953, 134-141
Robinson D.N. The Failure of the Coastal Defences of East Lincolnshire *Survey* 3, 2, 1953, 37-44
Robinson D.N. The Sea Defences of Lincolnshire *Lincolnshire Poacher* 1, 4, 1953, 29-32
Robinson D.N. Sea Surge Brought Disaster to Lincs. Coast *Grimsby Evening Telegraph* 29 January 1960
Robinson D.N. The Great Storm Flood of 1953 *Lincs. Life* 12, 11, 1973, 40-43
Summers D. *The East Coast Floods* (Newton Abbot 1978)
Conference on the North Sea Floods 31 January/1 February 1953—Institute of Civil Engineers 1954. Includes F.H. Tomes: Damage and Remedial Operations on the Lincolnshire Coast.
Lincolnshire River Board Report 1953

What shall we do with the Wash?
Doran W.E. The Wash *Geog. Mag.* April 1965, 885-897
Teggin H. City on the Wash *The Listener* 6 May 1966, 751-753
Teggin H. Britain's Europort: the real treasure in the Wash *Architects Journal* 15 January 1969, 142-198
White F. *A Basin Full of Water* (King's Lynn 1968)
Central Water Planning Unit: *The Wash Water Storage Scheme: Report on the Feasibility Study 1976*
Water Resources Board: *The Wash: Estuary Storage. Report on the Desk Study 1970*

Conserving the Coastline
Dugdale R.E. & King C.A.M. *A Guide to the Geomorphology of the Gibraltar Point Area, Lincolnshire* (Nottingham 1978)
Evans G. Gibraltar Point *in* Bird Observatories in Britain & Ireland ed. R. Durman (Berkhamsted 1976)
Gibbons E.J. *The Flora of Lincolnshire* (Lincoln 1975)
Kestner F.J.T. The Supply and Circulation of Silt in the Wash *I.A.H.R. Congress* (London) 1963, 231-238
Robinson D.N. Gibraltar Point Nature Reserve *Lincs. Life* 9, 6, 1969, 48-51
Robinson D.N. Humberston Fitties (unpub rept for Lincs. Branch CPRE 1966)
Robinson D.N. Humberston Fitties Holiday Camp: a brief on sea defence and condition of the foreshore (prepared for Grimsby RDC, 1972)
Robinson D.N. Samfer—a Lincolnshire Delicacy *Lincs. Life* 15, 9, 1975, 31-32
Robinson D.N. (ed) *Gibraltar Point Walkabout Guide* (Alford 1978)
Sheail J. The Impact of Recreation on the Coast. The Lindsey County Council (Sandhills) Act 1932 *Landscape Planning* 4, 1977, 53-72
Smith A.E. & Cornwallis R.K. *The Birds of Lincolnshire* (Lincoln 1955)
Smith A.E. *Nature Conservation in Lincolnshire* (Lincoln 1969)
Smith A.E. et al The Conservation of Nature in Lincolnshire *in* Nottingham and its Region ed. K.C. Edwards (Nottingham 1966)
Lindsey County Council *The Lindsey Coast: A Policy for Conservation* (Lincoln 1971) *The Lindsey Coast: A Policy for Holiday Development* (Lincoln 1973) *The Lindsey Coast: Access & Car Parking* (Lincoln 1974)
Lincolnshire County council *Development on the Lincolnshire Coast: Subject Plan* (Draft) (Lincoln 1981)
Nature Reserves Handbook 2nd ed.: Lincolnshire & S. Humberside Trust for Nature Conservation (Alford 1981)
UK Fortifications Club: World War Two Defences of Lincolnshire (unpub rept 1979)

General
County newspapers—particularly *Lincoln, Rutland & Stamford Mercury, Grimsby Evening Telegraph, Skegness News, Skegness Standard, Louth Advertiser.*
County Directories and a wide range of brochures and guides to Cleethorpes, Mablethorpe & Sutton, Chapel St Leonards, Ingoldmells and Skegness.

Key to Caption Credits

A	Aeropfilms
BK	Mrs B Kirkham
C	Cambridge University Collection
CES	C E Smith
DD	David Dunn
DL	Douglas Lamming
E	*The Engineer*
EHR	Mrs E H Rudkin
GET	*Grimsby Evening Telegraph*
M	G F Morton
HLC	History of Lincolnshire Committee
JS	J Stuart
JKStJ	J K St Joseph
JM	Jeffrey May
LCC	Lincolnshire County Council
LLS	Lincolnshire Library Services
LL	*Lincolnshire Life*
LNU	Lincolnshire Naturalists' Union
NS	Northgate Studio
RBW	R B Wilkinson
RF	Rev R. Fielden
RMR	Ruth Race
RWD	R W Dales
WERH	W E R Hallgarth Collection, Welholme Galleries, Grimsby
WK	Winston Kime

All others from the author's collection

Index

Figures in *italics* refer to illustrations

Subscribers

Presentation Copies

1 Lincolnshire County Council
2 University of Nottingham
3 Lincolnshire Library Service
4 Humberside Library Service
5 Lincolnshire & South Humberside Trust for Nature Conservation
6 Professor Emeritus K.C. Edwards CBE, MA, PhD, FRGS
7 A.E. Smith OBE, MA

8 David Robinson
9 Clive & Carolyn Birch
10 Mrs L. Robinson
11 Christopher D. Robinson
12 Hilary A. Robinson
13 Ralph H. Robinson
14 R. Brown
15 W.T. Day
16 K. Redmore
17 Mrs S. Turner
18 *Lincolnshire Life*
19 Anne K. Hawke
20 David Kaye
21 Mrs Cicely Irene Wilkinson
22 Rear Admiral G.P.D. Hall
23 A.A. Hayward
24 Mrs A.D. Stocks
25 Stuart Larder
26 George Bryan Hewson
27 George William Hewson
28 Mr & Mrs G.A. Todd
29 Miss Dorothy King
30 G. Hallion
31 Mrs S.M. Taylor
32, 33 Mrs J. Burton
34 Mrs J. Makinson-Sanders
35 Mrs Y.F.L. Heward
36 Peter Chard
37 David Read
38 Malcolm Read
39 Victoria & Albert Museum
40 Leicester Library
41 B.S. Kiek
42 J.N. Whitworth
43 Mrs W.L. Durham
44 Mrs A.M. Hunt
45 Mr & Mrs S.L. Josephs
46 Derek Norris Hunt
46 Miss N. Jenner
48 Miss Sheila D.J. Hannam
49 Mrs D.M. Brereton
50 Miss P.M. Gibson
51 Kenneth R. Dawson
52 Mrs P.J. Howard
53 D.J. Hunt
54 Mrs Jean M. Lincoln
55 W.J.S. & Mrs H.M. Grierson
56 A.G. Davies
57 Mr & Mrs A.R.Platts
58 Dr Andrew C. St M. Mills
59 E.H. Shelbourn
60 Mrs Rita Clapson
61 Pauline E. Thornton
62 John Laking
63 Roy Harrison
64 Mrs P.A. Lill
65 W.F. Sidley
66 James Moore
67 Mrs G. Swift
68 Dr Peter T. Oke
69 A.M. Barker
70 Mr & Mrs R.H. Grantham
71 Philip Grantham
72 Mr & Mrs R. Salton
73 J.W. Bickley
74 A. Grainger
75 Mrs J. Pridgeon
76 M.A. Ward
77 C.H. Lee
78 G. Gostick
79 Mrs M.E. Watson
80 David Cuppleditch
81 W.D. Dickinson
82 Eleanor Stamp
83 Mrs M.L. Williamson
84 Harold Jackson
85 Miss Pauline Robinson
86 Mrs W. Johnson
87 Mr & Mrs David John Harper
88 Mrs Grace Jacklin
89 Mr & Mrs J.R. Needham
90 Mr & Mrs G. Barry Sharpe
91 A.E. Jackson
92 C.J. Sturman
93 Christine Ruffell-Ward
94 Brian Ruffell-Ward
95 T.M. Briggs
96 K. Redmore
97 South East Lindsey Teachers Centre
98 Mr & Mrs Brian Brook
99 Dr & Mrs R.P.T. Hills
100 Mrs K. Ward
101 Mrs M. Collingwood
102 Ralph Sharp
103 Charles T. Smith
104 R.G. Jackson
105 Mrs P.A. Draper
106 Mrs C. Wilkinson
107 Mrs Mary Hudson
108 Neil R. Wright
109 Mrs C. Kitchen
110 Miss K.D. Johnson
111 Museum of
113 Lincolnshire Life
114 M.J. Roberts
115 J. Paul
116 L. Elvin
117 Mrs B. Webster
118 Dr C. Clubley
119 Lesley Colsell
120 Miss Jayne Knight
121 D.S. Jones
122 A.M. Taylor
123 Mrs M. Poole
124 Mrs J. Moore
125 Mrs D.A. Curtis
126 Mrs R.A.E. Dear
127 P. Morrell
128 Mr & Mrs N.J. Camamile
129 N.J. Camamile
130 Mr & Mrs A. Dowling
131 J. Barrett
132 Mrs O. Lakes
133 David John Griffiths
134 Mrs L.M. Wilson
135 C. Kemp
136 Mrs N. Jaines
137 Mrs Kathleen Bradshaw
138 Peter J. Shaw
139 R.W.Ambler
140 M.W. Ramsden
141 Michael M. Sleight
142 W.T. Ramsden
143 Harold Roberts
144 Patricia Anderson
145 Dr A.W. Smith
146 B.G.H. Roffey
147 A.P. Dixon
148 Dora Teanby
149 Mrs D. Kent
150 J. Oaten
151 S.P. Haagensen
152 D.A. Easton
153 Mrs D. Yardley
154 Mrs D.E. Sharlotte
155 R. Haith
156 Clive George
157 Mrs H. Odlin
158 John Zalewskyj
159 Graham Day
160 A. Ferrier
161 Mrs Thelma Hawkins
162 Frank Smith
163 Iain N. Reynard
164 Mrs E. Pettitt
165 R.W. Maidens
166 Mrs J.C. Ward
167 Mrs S. Portess
168 Mrs L.C. Wells
169 Mrs J.M. Blackburn
170 K.C. Denniss
171 R.T. Sharkey
172 Mrs S. Brydges
173 F.L. Parker
174, 175 M.J. Hughes
176 C. Moore
177 J.C. Mackness
178 Mrs P. Lill
179 W. Kine
180 H.W. Johnson
181 F.A. Forster
182 Mrs P. Severn
183 L. Smith
184 D.H. Plant
185 Miss G. Townsend
186 Mrs I. Peden
187 Mrs M.P. Burr
188 K.W. Needham
189 H.E. Spence
190 Mrs M.E. Muggeson
191 A. Kernahan
192 Mrs C.C. Morris
193 H.L. Clark
194 Mrs M. Wood
195 Mrs Joan Fordham
196 H.B. Bradley
197 Mrs D.M. Franklin
198 Mrs M.S. Barton
199 M.R. Chambers
200 Mrs A. Cox
201 R. Smeeton
202 George Boot
203 Dr O. Wright Holmes
204 Mrs Jane Richardson
205 Angela Margaret Packer
206 Mrs A. Stainforth
207 D.F. Wright
208 Miss M. Jenkinson
209 Colin Bark
210 Helen Gosling
211 R. Blackbourne
212 T.J. Netagh
213 S.O. Lindgren
214 K.B. Collinson
215 Roy B. Turrell
216 Robin Leak

217 Margaret A. French
218 Mrs Key
219 R.W. Dales
220 H. Culme
221 Valerie Ann Gosling
222 R.B. Lindley
223 Mrs M. Bark
224 H.C. Tester
225 Mrs B. Coupland
226 Mrs M. Roper
227 B. Buddle
228 Jane Murphy
229 R. Farrow
230 Mrs M. Clarke
231 Mrs Elizabeth Varney
232 D.J. Brailsford
233 Mr Blackburn
234 P. Chambers
235 Mrs D. Mountain
236 L. Stretton
237 F.W. Cross
238 G.W. Baxter
239 T.E. Merriman
240 Roy Allen
241 J.A. Bayles
242 David K. King
243 Carol Shennan
244 / 245 Mrs May Bell
246 Mr & Mrs R.J. Fenwick
247 / 250 T.R. Sutton
251 Michael May
252 R. Near
253 E.H. White
254 / 255 Mrs N. Tuff
256 / 257 Miss J.M. Watson
258 Mrs E. Haynes
259 Alex G. Parker
260 Derek William Thorpe
261 / 262 Christopher Dobson
263 B. Mather
264 D.J. Codling
265 / 266 Dr J.A. Galletly
267 George E. Cotsworth
268 Dr Peter Bush
269 D.S. Whicher
270 Rex Critchlow
271 Mrs J. Quinn
272 C.R.M. Dowson
273 Graham Couling
274 Mrs E.L. Thornley
275 E.G. Martin
276 Olive Davis
277 D.S. Goldsbrough
278 Councillor Miss L. Parkes
279 Mrs A. Robinson
280 Dorothy Swinn
281 Albert E. Davie
282 Roy & Sandra Davie
283 Mrs E.H. Rudkin
284 Mrs Patricia Skudder
285 Mrs Barbara Beels
286 Norman L. Brown
287 Miss Pauline Johnstone
288 / 289 Leas Junior School
290 Doris Croft
291 D.W. Parr
292 C. Newton
293 John J. Rann
294 Mrs A. Benson
295 Mrs J. K. Lowe
296 Mrs Patricia Jowes
297 / 298 Miss E. Miller
299 Mr & Mrs Norman Heath
300 Michael John Heath
301 Richard Timothy Wallis
302 Peter Alwyn Sydney John Axe
303 Robin James Harry Sumpter
304 Alan Everett Jackson
305 A.J. Hutchings
306 Mrs Leola Kneebone
307 M.I. Seal
308 / 309 F.A. Everett
310 Clifford Morrison
311 R. Broughton
312 Mrs M. Winter
313 Sleaford County Secondary School
314 Rodney Easton
315 Mrs M. Pounder
316 Mrs Peggy Francis
317 Denis Robson
318 R. Watson
319 Boyce Smith
320 Rosamund Higginson
321 C.B. Stansfield
322 Mrs B. Mumford
323 Joyce Skinner
324 Barbara Snowden
325 E.J. Garlick
326 Mrs E.B. Topley
327 J. Aldridge
328 John Hamilton
329 / 330 Canon J.E. Swaby
331 Dennis William Harris
332 David J. Wedd
333 T.R. Wilson
334 Cordeaux High School
335 J.M. Walker
336 S.J. Cooper
337 P.C. Dibben
338 Miss Jessie Skellern
339 Geoffrey A. Young
340 Dr Allan H. Briggs
341 Rosemary Heath
342 Kidgate C.P. School
343 Mrs D.P. Barron
344 Mrs M. Wyers
345 Mrs A.B. Lacy
346 Miss E. Smith
347 T.H. Thompson
348 M.H. Duncalfe
349 Mrs B.M. Barrett
350 G.R. Oxby
351 E.J. Lucking
352 Mrs D.R. Patman
353 A.J. Jones
354 Catherine Kyle
355 A. John Rhodes
356 Neil Pearson
357 Claudia Kirk
358 Dr D. Nigel Lock
359 R.B. Lindley
360 Timothy J. Hudson
361 Leicestershire County Council
362 James Henry Foster
363 Douglas B. Smith
364 D. Turner
365 E.W. Roberts
366 Mrs A.E. Farrow
367 Cliffedale School
368 J.E. Smith
369 Mary Large
370 W.M. Peet
371 Ingoldmells C.P. School
372 Chapel St Leonards C.P. School
373 Ronald Morter
374 Queen Elizabeth Grammar School, Horncastle
375 Pamela McGill
376 Barbara Thomas
377 H.B. Banks
378 J. Morris
379 C. Stamp
380 F.G. White
381 P. Coote
382 D. Hardcastle
383 Jean Stovin
384 John Russell Ingamels
385 Miss E. Hole
386 Brian Watton
387 Mark R. Giles
388 Leslie Hillard
389 Joan V. Finch
390 Henry Walter Gardam
391 R.A. Swingen
392 Boston Tower Road School
393 R.V. Sheppard
394 A. Ward
395 Miss M.A. Toomer
396 N.A.N. Sharplby
397 Dr D.L. Richards
398 Spalding Gentlemens Society
399 Maurice Rutherford
400 The Rev D.F. Tunbridge BA
401 / 405 O. Raybould
406 J.S. Robinson
407 P.J. Rhodes
408 L.W. Scarfe
409 R.A. Scarfe
410 Ms I.P. Heward
411 J.W. Hardy
412 D. Belton
413 H.J. Atkinson
414 John Ingall
415 J.G. Lomas
416 Mrs Susan Weaver
417 Miss J. Whittaker
418 Derek C. Grant
419 John Appleby
420 Mrs J.L. Smeeton
421 L.W. Ronson
422 Neville Mudd
423 R.P. Harrison
424 John R. Barker
425 Margaret Cox
426 H.C. Wilkinson
427 B.C.V. Spence
428 F.A. Jackson
429 R. Kenneth Day
430 S. Smith
431 / 433 Miss D.M. Dixon
434 Peter C. Robinson
435 Mrs Joy Wilkinson
436 Mrs M.G. Burgoyne
437 Mrs C.M. Dawson
438 F.S. Waddington
439 Eddie Squires
440 B.L. Leeke
441 Fred A. Cant
442 Mrs E.M. Stamforth
443 Peter K. Dennis
444 Mrs B.M. Smith
445 Mrs J. Smith
446 / 447 John M. Keily
448 S.E. Crooks
449 M.A. Cross
450 Peter W. Desforges
451 Mrs E.N. Bullock
452 Mrs M.R Giles
453 R.W. Larder
454 Rex & Jenifer Critchlow
455 R.L. Done
456 P.J. Smith
457 W.G.M. Dickinson
458 C.T. Herbert
459 Mrs J. Barnard
460 Rev John G. White
461 N.E. Whitaker
462 Miss S. Mackenzie
463 Gwenda V. Bell

464 Anthony John Dent
465 Donovan N. Fry
466 J. H. Portas
467 David B. Griffiths
468 Charlotte Alderton
469 R.A.O. Hickling
470 Miss M. Gilyead
471 Peter Hargreaves
472 Robert Henry Crawford
473 K.C. Johnson
474 G.R.S. Pepler
475 Alan Geoffrey Whalley
476 R. Overton BA
477 Rev C.H. Overton
478 Dr David Crook
479 480 M.R. Curry
481 Irene Lingard
482 A.G. Burdett
483 Patrick Cormack FSA, MP
484 Andrew L. King
485 H.R.B. Marwood
486 H. Lester
487 Norman Leveritt
488 Paul Empringham
489 Audrey Oppice
490 John R. Anthony
491 G. Couling
492 Andrew L. King
493 David Brown
494 Mrs J. Pearson
495 496 Mrs J. Overton
497 B. M. Jackson
498 Mrs Trimble
499 P.R. Garner
500 Lincolnshire Museums
501 F. Smith
502 Mrs I. M. Bailey
503 Dr D.S. Hill
504 Mrs Irene Lingard
505 Mrs Anne E. Ward
506 Mrs A.M. Pearson
507 Mrs D. Price
508 R. Yorston
509 R. J. Wright
510 B.R.T.Hopwood
511 David Swift
512 Miss Florence Brader
513 B.R.T.Hopwood
514 M.J. Jackson
515 John Gray
516 G. Kirby
517 Mrs K.M. Tuplin
518 Frederick Marshall
519 Mrs M. Bartle
520 H. Lomas
521 John Mitcham
522 H.E. Brocksom
523 Miss N. Pridmore
524 Jean Lillie
525 Mrs J. Graville
526 Mrs A.D.M. Wainer
527 Lucy Eileen Cussons
528 Mrs P. James
529 John E. Hall
530 Mrs H.W. Haseldine
531 Edward H. Trevitt
532 Miss F.C.M. Hay
533 P.F. Fleet BA
534 University of Nottingham
535 Dr J.D.R. Mawer
536 Anton Rippon
537 W.P. Roe
538 Miss Anne Topliss
539 Monks Dyke High School
540 Mrs Pamela Berry
541 E.R. Coddington
542 Alan Tye
543 J.A. East
544 W.B. Hatherley
545 Rodney E. Callow
546 D.G. Boulton
547 Julie A. Vint
548 Miss M.D. Rostron
549 B.R.T. Hopwood
550 Pilgrim College
551 Frank Crowder
552 Ronald Drury
553 Terence R. Leach
554 Kitwood Boys School
555 Dr Harry Jackson
556 J.A. Stubbs
557 Christopher C. Hunt BA
558 Douglas Lamming
559 F.T. Baker OBE, FSA
560 Mr & Mrs C.R. Fanthorpe
561 J.N. Clarke
562 Martin Curry
563 Olive L. Grosvenor
564 Miss K.D. Venables
565 Michael & Diana Honeybone
566 Rita J. Rhodes
567 Janice A. Montgomery
568 John F. Bailey
569 J.T. Turner
570 Eileen Robson
571 Eric S. Smith
572 P.B. Grimes
573 Willoughby Memorial Trust
574 Malcolm G. Knapp
575 576 Rosemary Oliver
577 George H. Parker
578 R.M. Taylor
579 Mr & Mrs I.R. Walters
580 Miss P. Wheatley
581 Alan Clark
582 C.J. Lester
583 Ethel H. Rudkin
584 Miss M.E. Maltby
585 586 A.E. Smith
587 Mrs Frances L. Anderson
588 Mrs Betty Kirkham
589 Belinda Barratt
590 C.J. Ketchell
591 J.S. English
592 601 Lincolnshire Library Service School Library Fund
602 691 Lincolnshire Library Service
692 Freiston Hall Field Centre
693 John D. Lawrence
694 Malcolm Pool
695 Mrs J.G. Sardeson
696 697 Bishop Grosseteste College
698 K. Hollamby
699 Andrew L. King
700 Miss M.M. Drewery
701 A.E.B.Owen
702 C.R. Brears
703 Mr & Mrs J.B. Hammerton
704 705 Robert A. Frankum
706 David Dunn
707 North Lindsey College of Technology
708 John Keith Shaw
709 Mrs S. M. Sharp
710 W.A. Cooling
711 M.G. Johnson
712 Alan Beesley
713 Lieutenant Colonel M. Proctor
714 A.W. Reynolds
715 Ruth Toulson
716 Mrs R.M. Rowe
717 H.M. Hipkin
718 Chris & Mabel Jackson
719 David & Sheila Greenfield
720 Sir Peter Kent
721 M.E. Congreve
722 Miss L. Jekyll
723 D.E. Woolhouse
724 Mrs J. Rowling
725 Boston College of F.E.
726 P.J. Blakeborough
727 R.W. Taylor
728 K. Tovey
729 R.J. Worraker
730 R.A. Rogerson
731 P. Stanbridge
732 C.I. Smith
733 A. Heath
734 Dr Peter Hurst
735 R. Pearson
736 T.W. Taylor
737 Mrs P.E. Pomeroy
738 B.M. Shelbourn
739 Mrs H. Watton
740 S.P. Ellis
741 Geoff Rawlings
742 Mrs E. Ellis
743 Miss S.G. Levy
744 Mrs L.E. Anderson
745 P.E. Gilbert
746 Mrs W. Smith
747 John M. Enderby
748 J. Egar
749 Dr A.J.H. Giles
750 Mrs J.B. Hill
751 P.F. Place
752 Mrs M.M. Smart
753 N. Russell
754 John Topley
755 Mrs J. Borrill
756 D.F. Fletcher
757 N.P. Hayselden
758 Mrs H. Hadwick
759 P.G. Byllam-Barnes
760 Lindsey School Library
761 A.K. Taylor
762 Scothern C.E. School
763 Mrs Marjorie Draper
764 Mrs M.E. Sayles
765 Mrs B. Kirkham
766 Marjorie & Jim Seal
767 Marjorie Blatherwick
768 Peter Hergreaves
769 Miss K. Showler
770 771 Miss E. Hole
772 T.J. McVeigh
773 Don & Joy Garwood
774 803 Humberside County Council
804 Mrs Margaret Hewson
805 D.J. Gale
806 Alan Daniels
807 S.O. Lindgren
808 Mrs R. Keilly
809 810 University of Nottingham Dept of Adult Education

Remaining names unlisted

BACK COVER: The famous John Hassall
painting of the jolly fisherman of Skegness.

ENDPAPERS: The Lincolnshire Coast from Armstrong's map of 1779.

Quadring
Strangle
Alms House
Binthorpe
Billingborough
Stow Green
Monkhouse
Ramsey
Sutterton
Gosberton
London 107 Miles
Westhorpe
Sempringham
Algarkirk
Fosdike
Dowgate
Old
Fosdike
Swine Green
Pointon
Clow
Risegate
Eau
Gresey Hall
A N D
Barr
Belby
Hall
Rigbolt
Pinchbeck
Banks
Dowsby
Milthorpe
Chale
Reservoir
Rippingale
Surfleet
Wrag Marsh
and Spalding
Road to Donington
Dunsby
Six Houses
Ogle Hall
Sea
Common
Ruins
Old
Vernattass Drain
Heckingby
Dovehirn
Barrs
Pinchbeck Hall
Pinchbeck
Clowhirn
Morton
GLEN RIVER
Hill
Wykeham Chap.
HOL
Dyke
Mill Green
Weston
Fondly
Barr
South
End
Gutthrem
Spalding
Podge Hole
Monks House
SPALDING
Race
ground
100 M. from London
Bourn River
Bourn
Cowbit
Tongue End
Counter Drain
Road from Sp
Commons
North
Drove
Moulton Chap.
Cowbit
Thurlby
Peakhill
Adventurers Lands
Adventurers Lands
South Drove
Wilsthorpe
Kates Bridge
The Ford
Sheep Wash
Old Welland
Tarketale Cross
Brother House
Baston
Wash
Cloot
White House
Crowland Common
or
Goggusbland
Boundary
Barr
of
Holland
Great Drove
Porsand
James
Welland New Navigation
Crowland
Langtoft
E S S
Sheep Skin Hall
Stow
CROWLAND
Ruins
Dowsdale
Barrs
Cranmore
Frognall
Common
Hersgate
James Deeping
St. Kauulphs Cross
St. Guthalac
MARKET DEEPING
West Deeping
Barr
Crowland Common
Blackhorse
New South
Waldram Hall
Pepper Mill
Turketals Cross
CAMBR